Japanese Whaling and the People Behind It

A Look from Within

This book explores the recent developments in global and Japanese whaling from the viewpoint of the members of the Japanese whaling community, a perspective that is largely neglected and misinterpreted.

Japanese whaling has been one of the most contentious issues in global environmental governance in recent years, and Japan is often harshly criticized for its whaling programs. By distinguishing between the different whaling-related actors and their experiences, this book widens our understanding of why whaling programs continue to exist. Rich in ethnographic data, the book includes in-depth interviews with representatives of the Japanese whaling community, from government officials to fishermen, shedding light on what whaling represents, both historically and today.

As an ethnographic study of a divisive and controversial subject, this book will appeal to a wide range of students and scholars, including, but not limited to, those interested in Japanese studies, anthropology, political science, and ocean resource management.

Nadzeya Shutava is a social science researcher residing and working in Washington, D.C., USA. She holds a Ph.D. in International and Advanced Japanese Studies from the University of Tsukuba, Japan.

Routledge Contemporary Japan Series

For more information about this series, please visit: *www.routledge.com/Routledge-Contemporary-Japan-Series/book-series/SE0002*

Japanese Whaling and the People Behind It

A Look from Within

Nadzeya Shutava

Routledge
Taylor & Francis Group

LONDON AND NEW YORK

First published 2024
by Routledge
4 Park Square, Milton Park, Abingdon, Oxon OX14 4RN

and by Routledge
605 Third Avenue, New York, NY 10158

Routledge is an imprint of the Taylor & Francis Group, an informa business

British Library Cataloguing-in-Publication Data
A catalogue record for this book is available from the British Library

ISBN: 978-1-032-18535-4 (hbk)
ISBN: 978-1-032-18536-1 (pbk)
ISBN: 978-1-003-25503-1 (ebk)

DOI: 10.4324/9781003255031

Typeset in Times New Roman
by KnowledgeWorks Global Ltd.

Contents

Figures

Acknowledgments

I would like to express gratitude to my academic supervisor, Dr. Timur Dadabaev. The conversations on the meaning and value of academic work in today's world we had during my years at the University of Tsukuba helped me shape the vision of what and how I wanted to contribute to social science research. Thank you for believing in me and for letting me follow my curiosity instead of choosing a more secure path.

I am also thankful for the comments I have received from other professors at the University of Tsukuba. Dr. Leslie Tkach-Kawasaki, Dr. Akifumi Shioya, and Dr. Nathan Gilbert Quimpo, the dedication you have for your work and the enthusiasm you treat your students' research projects with are truly inspirational.

This book would not have been possible without the trust and support I received from the people involved in the issue of Japanese whaling, both in Japan and outside of the country. Joji Morishita, Hideki Moronuki, Kiyokazu Yoshimura, Hirohiko Shimizu, Yuriko Shiraishi, Akiko Sato, Gavin Carter, and many others – I will never be able to thank you enough for being patient, open-hearted, and open-minded with me and my ideas. Thanks to you, I was able to be present in places I was not supposed to be present at and witness first-hand the processes behind the Japanese whaling policies, as well as whaling-related practices that not many "outsiders" are allowed to witness. It was thanks to these warm welcomes and unique experiences that I was able to continue my research with passion and dedication for what in today's fast-paced world is considered an eternity – four remarkable and memorable years in Japan and several more years post-Ph.D. to make this book happen.

All the participants of this research had no incentive to share their time and energy with me other than their own generosity and respect for other people's work. I am forever grateful for encountering people like this on my personal and professional journey – to this day, this inspires me to mentor and share my knowledge and skills whenever I can.

I was incredibly lucky to have the emotional and practical support of my Tsukuba University colleague and friend, Roksolana Lavrinenko. Knowing that you have someone you can always rely on is crucial to the quality of one's day-to-day life and work. Academia is not an easy environment to navigate, and I believe that as its representatives, we need to be kinder and more supportive of one another, guided by the simple idea that we are in it first and foremost for the common goal of new discoveries.

It's also important for me to acknowledge you, my reader. Whether you know a lot about whaling in general and Japanese whaling in particular and have a strong opinion on it, or picked up this book because you have an interest in anthropology, or Asian studies, Japanese culture, or because whales are your favorite animals – I had you in mind while writing this, and I am grateful for your intellectual curiosity. I also thank you in advance for engaging in intentional perspective-taking – a vital life skill of considering the situation from the perspective of another person, taking into account their specific circumstances that may significantly differ from your own – this is of utmost importance to the understanding of this book and its characters.

Finally, mom, dad, and husband – I was on an emotional roller coaster while working on this project, and you braved it all with grace. As a result, we have this book — because, honestly, it's yours too — and a wealth of internal jokes about it. I love you forever.

Glossary

Aboriginal Subsistence Whaling (ASW)   ASW is a type of whaling that is practiced by aboriginal communities where, according to a definition by the International Whaling Commission, "whale products play an important role in the nutritional and cultural life of native peoples"[1]. Currently AWS is officially conducted by communities in four countries – Denmark (Greenland), Russia (Chukotka), St Vincent and the Grenadines (Bequia) and the United States (Alaska).

Baleen   An apparatus inside baleen whales' mouths consisting of up to 900 plates hanging in a row from the animals' upper jaw. Baleen is used as a filter-feeder system. Baleen whales have their mouths open as they swim to allow large amounts of water along with krill and fish to flow into their mouths. The water then flows out through the gaps between baleen plates, while krill and fish are filtered to stay in and are swallowed afterwards. Baleen plates differ in length and other properties depending on the whale species, and the exact feeding methods vary slightly as well.

Baleen, also called "whalebone" and dubbed "the plastic of the 1980s", was one of the key products of interest for the whaling industry in the 19th–20th centuries. Due to its strength and flexibility, baleen was used in manufacturing of corsets, collar stays, whips, toys, umbrella ribs, fishing rods, springs in typewriters, carriage springs, etc.

Blubber   A thick dense layer of fat (adipose tissue) under the skin of cetaceans and other marine mammals. Covers the whole body of the animals except for appendages.

Blubber was the principal product of interest of the whaling industry in the 19th–20th centuries, as it was used for rendering into oil. The oil in its turn was used as fuel, as well as in manufacturing of cosmetics products, margarine, candles, soap, machinery lubricants, paint, varnish, etc.

Cetaceans   Cetaceans (order Cetacea) are an aquatic order of mammals that includes whales, dolphins and porpoises.

Exclusive Economic Zone (EEZ)   In international law (see United Nations Convention on the Law of the Sea, Part V) an Exclusive Economic Zone or EEZ is "an area beyond and adjacent to the territorial sea" of a state where that state has jurisdiction over resources. It extends up to 200 nautical miles (370 km) from a country's land coasts.

Factory ship A type of large ocean-going vessel that is constructed to include whale- or fish-processing and freezing facilities and equipment on board. Some factory ships are also **mother ships** – they carry smaller catcher ships/boats on board.

The whaling industry is where the idea of factory ships originated - they were designed to make whaling trips more efficient by reducing the number of times the ship had to return to its port. The concept was later adapted to the needs of fishing and is still successfully used.

Flensing Slicing the skin and blubber of off a carcass, often a whale's carcass. Some texts use "flensing" referring to removing meat from a carcass and cutting it up as well. The Japanese terms for flensing are 魚切 *"uo-kiri"* (lit. fish cutting) or 解体 *"kaitai"* (lit. dismantling, disassembly). Flenser is a specialist who performs the task of flensing.

High seas In international law high seas are "all parts of the sea that are not included in the exclusive economic zone, in the territorial sea or in the internal waters of a State, or in the archipelagic waters of an archipelagic State" (United Nations Convention on the Law of the Sea, Article 86). High seas are open for all states to navigate and use, including for fishing, research, construction, etc.

Large-type coastal whaling (LTCW) LTCW and STCW (see below) are two categories of whaling that Japan recognizes for itself.

In LTCW in most cases large vessels are used and larger species are targeted, such as baleen whales that are larger than minke and sperm whales. LTCW was paused in 1986–2018 in accordance with the requirements of the International Whaling Commission moratorium on commercial whaling. It was restarted in 2019 after Japan withdrew from the International Convention on the Regulation of Whaling (ICRW) and the IWC.

Small-type coastal whaling (STCW) STCW and LTCW (see above) are two categories of whaling that Japan recognizes for itself.

In SCTW only certain types of smaller-sized whales and dolphins are caught – mainly minke whales (the hunts for minke whales were suspended while Japan was a member of the International Whaling Commission). Only vessels of 49 tons and below are used in STCW (compare to the size of Nishin Maru – a Japanese whaling mother ship of 8145 tons that was used in Antarctic whaling; or Yushin Maru – 724 tons, Yushin Maru No. 2 – 747 tons and Yushin Maru No. 3 – 742 tons – all Japanese whaling vessels used in pelagic whaling).

The Japanese delegation to the IWC made attempts to use the category of SCTW during international negotiations, mainly to advocate for different regulations of catch for Japan's STCW communities. However, the category itself, as well as the requests for treatment of STCW any differently than commercial whaling were repeatedly rejected by the IWC. During 1990–2018 Japan submitted proposals for STCW quotas at the IWC Committee meetings 21 times.

Note

1 See more information on this type of whaling online https://iwc.int/management-and-conservation/whaling/aboriginal

Abbreviations

ASW	Aboriginal Subsistence Whaling
BWU	Blue Whale Unit
CITES	Convention on International Trade in Endangered Species of Fauna and Flora
ENGO	environmental non-governmental organization
FAO	Food and Agriculture Organization
HIS	Humane Society International
ICJ	International Court of Justice
ICR	Institute of Cetacean Research
ICRW	International Convention for the Regulation of Whaling
IFAW	International Fund for Animal Welfare
IGO	intergovernmental organization
IKAN	Iruka and Kujira Action Network
IWC	International Whaling Commission
JARPA	Japanese Research Program in the Antarctic
JARPN	Japanese Research Program in the North Pacific
JWA	Japan Whaling Association
LTCW	Large-type coastal whaling
MAFF	Ministry of Agriculture, Forestry and Fisheries (Japan)
MMPA	Marine Mammal Protection Act
MOFA	Ministry of Foreign Affairs (Japan)
NAMMCO	North Atlantic Marine Mammal Commission
NEWREP-A	New Scientific Research Program in the Antarctic Ocean
NMP	New Management Procedure
RMP	Revised Management Procedure
SC	Scientific Committee (IWC)
SOS	Southern Ocean Sanctuary
STCW	Small type coastal whaling
UNCLOS	United Nations Convention on the Law of the Sea
WAO	Whaling Affairs Office (MAFF)
WFF	Women's Forum for Fish
WWF	World Wildlife Fund

Introduction

During the seven years this project lasted, people with whom I had a chance to discuss it often wondered how and why I initially took an interest in whaling. This question would be asked regardless of the setting, usually with a considerable degree of surprise. Participants of academic conferences would be wondering just as much as those at leisure events with attendees from diverse backgrounds. Many of the participants of this study were also curious about how this project started since I originally had no obvious connections to Japanese whaling and the Japanese whaling community.

During the first year of my doctoral program, in 2015, I was working on narrowing down my interest in maritime security. At some point, I came across a CNN article about Japan "defying the world" and restarting its research whaling in the Antarctic Ocean regardless of the International Court of Justice (ICJ) ruling against it (Whiteman, 2015). Although I had already been studying the Japanese language and society for more than ten years by then, I had not known much about whaling, had never consumed whale products or known if anyone had, and had not been aware that whaling was a heated international issue provoking protests and even violence. That is despite the fact that Japan is one of the few countries in the world that still engage in whaling and where you can find some whale-meat-based cuisine.

This short article on Japanese whaling caught my attention, as it contradicted the image of "a good global citizen" Japan is actively constructing by demonstrating, among other things, its commitment to environmental protection. Some authors debate whether Japan is actually doing well in terms of eco-mindfulness,[1] but according to the country's official statements, Japan has aspirations of spearheading the world's efforts to slow down and reverse the climate crisis (MOFA, 2018). Bringing up the example of the Kyoto Protocol[2] negotiations, Kolmaš (2017) argues that Japan uses environmental multilateralism to strengthen its own identity as an independent and proactive state on the global stage. In this context, it is unexpected to read about Japan "defying" the rest of the world to "cruelly" "kill" an "endangered species" without an immediately understandable reason behind such actions.[3] My surprise reading these words was how this research started and with time became a quest in search of other perspectives, accurate details and personal stories of Japanese people involved in whaling. All of these, in their turn, "defy" the one-dimensional account of whaling as inexplicable cruelty on the part of Japanese people.

DOI: 10.4324/9781003255031-1

Japanese whaling issue dynamics

Chapter 1 of this book gets further into the global history of whaling, the details of how it developed specifically in Japan, and the important role history plays in the pro- and anti-whaling debate. But to put things into perspective the gist of the history and the present day of whaling is outlined below.

The history of whaling starts thousands of years ago, way before the history of the whaling and anti-whaling debate. It was only in the 20th century that this activity started to cause controversy. The trends in whaling before 1982 can be summed up in a few key words that mirror the developments in the overuse of global resources in general – increasing consumption, fierce competition, and over-harvesting. During most of the 20th century countries all over the world were failing to regulate whaling activities – for themselves and collectively. This resulted in the severe depletion of stocks for many whale species. After 1982, however, because of the industry's diminishing returns and thanks to environmental activism, whaling as a global industry de facto ceased to exist.

In 1982, the International Whaling Commission (the IWC) paused all commercial whaling.[4] The decision came after it was "discussed extensively" and one of the reasons for it as stated in the Chairman's Report of the Thirty-Fourth Annual Meeting of the IWC (1982) was the absence of reliable data on sustainable harvesting. This "pause", also referred to as a "blanket moratorium", was protested by several IWC member countries, including Japan, South Korea, and Iceland. These countries believed that there was no scientific justification for pausing commercial whaling for all whale species and stocks since some of them were reliably proven to be healthy enough for sustainable hunts. Japan believed that the moratorium "violates the Convention as well as infringes sovereign rights in coastal waters" (Chairman's Report of the Thirty-Fourth Annual Meeting, 1982). Initially, the moratorium was introduced as a temporary measure and was supposed to be reviewed in five years (Chairman's Report of the Thirty-Fourth Annual Meeting, 1982). However, this never came to be and in 2023 the moratorium is still in place regardless of the increasing scientific proof of multiple whale stocks' availability for sustainable harvesting.

Japanese whaling has been one of the most contentious issues in global environmental governance during the past four decades, attracting international criticism and heated debates (Gales, Kasuya, Clapham, & Brownell, 2005). Soon after its initial objection to the moratorium Japan withdrew it in the face of threats of economic sanctions from the United States and agreed to stop commercial whaling by March 1988. At almost the same time Japan initiated its first research whaling program called JARPA (Japanese Research Program in the Antarctic), which was followed by the second reviewed phase of the program – JARPA II, and later on by NEWREP-A (New Scientific Research Program in the Antarctic Ocean), as well as JARPN (Japanese Research Program in the North West Pacific) and JARPN II. These research programs were criticized, the use of lethal methods in them questioned, as well as Japan's selling of research by-products (whale meat) in the country's markets. Many pieces of writing use "commercial whaling in disguise"

Figure 0.1 Anti-whaling protesters during the 67th IWC Committee Meeting in Floria-
noloppis, Brazil, September 2018. Photograph by the author.

referring to Japan's whaling research programs (BBC, 2015; Lies, 2014; Talmazan,
2019), and anti-whaling activists often feature these problematic, in their opinion, research programs in their speeches (personal observations, September 2018,
Brazil).

There were several international motions for Japan to stop its research whaling
programs, including a case filed with the ICJ by Australia with New Zealand intervening in 2010–2014. But until the 2018–2019 season Japan continued conducting
them.

A change of historic importance occurred on the December 26, 2018, when
Japan announced its withdrawal from the International Convention for the Regulation of Whaling (ICRW), which subsequently also meant the country's withdrawal
from the IWC – as the organization was established by the ICRW. Although this
move was a long time in the making, the final confirmation came as a surprise
to most in the international community. Japan's decision provoked a new wave
of publications on the issue, many of which time and again question why Japan
behaves the way it does when it comes to whaling.

According to the IWC's regulations Japan officially stopped being a member
on June 31, 2019. It might take several more years to be able to fully grasp the
direction(s) and scale of the shift this resulted in, but it is clear that a new period in
Japanese whaling is starting to slowly emerge.

While there are more details characterizing the problem, and the purpose of this
book is exactly to show how multi-faceted and complex it is, the presented above
overview is what a reader with a mild interest in the issue would understand from
doing a little online research and browsing through English-language resources.
An important additional detail to note here is that the readily available materials on

the subject can rarely claim neutrality – most have an obvious anti- or pro-whaling agenda. Another remarkable point is that almost none of the sources that such a somewhat curious reader would come across, explain what the people involved in Japanese whaling think about all of the above. One can expect to see "Why Japan will not give up whaling" (Danaher, 2002), "Japan approves whale meat import" (BBC, 2008), "Japan vows to carry on whaling" (ABC, 2009), "Japan prepares to resume commercial whaling" (Lewis, 2019) and "Japan wants to catch whales" (Lauer, 2020). But who and what exactly stands behind the notion of Japan in the context of the whaling controversy? This is a question that gets rarely asked and even more rarely answered and one we will dive into in this book.

All of the above brings us to the goal of the research behind this book – to show the issue of whaling from the viewpoint of the Japanese people behind it, giving them back the control over the narrative of their own experiences in the context of the major fluctuations happening in the perception and practice of whaling in the past several decades. The stories shared by chosen individuals and their activities as observed during the fieldwork done for this book, are collected, interpreted, analyzed, and put into context to form a holistic picture of Japanese whaling.

It is the premise of this study that the Japanese facet of the whaling issue has not been fully considered and understood in academia, popular media, and popular minds, and, subsequently, political debates. In the Western discourse, Japanese whaling is often seen and treated as a foreign oddity with the abstract "Japan" behind it. The distinction between different whaling-related actors is rarely made and understanding of their varying experiences is often lacking, which brings about confusion and misunderstandings. This book attempts to break down this image and show Japanese whaling as a puzzle, where every piece, be it involved in it subgroup of individuals, space they occupy, an event they organize, or relations between several such groups, has its designated part in shaping the whole.

Without understanding the experiences of people who shape the policies on whaling, or those who have been directly involved in this practice on the ground (or at sea!), it is easy to overlook the underlying reasons for the whaling controversy. To address this, we will dive into two overarching questions hereafter:

1   What characterizes the emic[5] (or internal, from within) side of Japanese whaling – what kind of first-hand experiences people engaging with the issue of Japanese whaling have been having?
2   What is the relationship between the emic (internal) perspective on Japanese whaling and the etic (external) perspective? How do these differ?

How the research was done

Among the various paths that could help answer the above questions a methodological approach inspired by the combination of phenomenology and ethnography was taken. Both phenomenology and ethnography often use interviews to capture first-person perspectives. The research done for this book included semi-structured in-depth interviews with 46 participants from different parts of Japan and beyond, all of them having a different kind of involvement with Japanese whaling.

Both phenomenology and ethnography also place routine life, the ordinary (for the participants, but in a lot of cases not for the rest of the world) at the forefront of their inquiry. To better understand the "ordinary" from the perspective of the Japanese whaling community, the data collected for this book also came from observations, casual interviews, and group discussions during more or less formal events and informal gatherings. I also reviewed pamphlets and reports shared by research participants and their affiliated organizations, of which there were many.

The Japanese "whaling events" that are mentioned throughout this book include a wide range of activities intended for the promotion of whale meat-based cuisine, explaining and promoting the Japanese whaling culture and Japanese whaling policies, etc. Attending these was especially valuable in understanding how whaling-related activities are woven into the daily lives of people whose lives are strongly connected to this practice, as well as the lives of those who have little to no relation to whaling.

While working on this book I also had the unique opportunity to attend the 67th International Whaling Commission (IWC67) Committee meeting, convened in Brazil in September 2018. This meeting can also be considered one such "whaling event", albeit international and of a much bigger scale than those mentioned above.

Figure 0.2 Sugamo Whaling Festival, where activities included a food market with whale products and dishes and a whale-themed poetry contest among others. Toshima, Tokyo, October 2018. Photograph by the author.

Figure 0.3 Whale soup distributed at Sugamo Whale Festival. Toshima, Tokyo, October 2018. Photograph by the author.

The IWC meetings are not open to the public – only government representatives, accredited media representatives, and registered NGO representatives are allowed to be present at the venue. They are also currently only held biennially. The restrictions on attendance and the timing were factors that made attending IWC67 look difficult when I started working on this project. But thanks to plausible circumstances and, more importantly, the trust and support of my research participants, this key component of fieldwork was added to the methodological arsenal of this project. This provided access to more perspectives, as well as the opportunity to verify the data that was collected through other sources. IWC67 was a stage where the theater of the global whaling issue held its regular performance. By the time I had a chance to witness this play, very little improvisation was involved – everyone's roles were well-rehearsed. But as such it was a fascinating "show" to watch in the context of the present research.

Both ethnographic and phenomenological approaches are also often used to create a holistic picture of a whole – by either collecting descriptions of its elements or by bringing to light "essences" that characterize many experiences of the same phenomenon. While ethnography helps to understand a group and its certain accepted "rules of engagement", phenomenology focuses on a phenomenon through individual accounts of it. Both these methodological or some would say philosophical, components of this research do justice to its goal – to see what Japanese whaling (a phenomenon) is like through the eyes of those who are directly involved in it (a group/community of people).

The whaling-related world of Japan is, using phenomenological terminology, a particular "lifeworld" that has its features and rules making it qualitatively different from the "lifeworlds" experienced by people whose lives are not related to whaling. Examining the different aspects of this lifeworld, this book centers on the perspectives of people whose positions have often been ignored and whose actions misunderstood. This book is intended to deepen the readers' understanding of the recent developments in global and Japanese whaling showing them through the eyes of the members of the Japanese whaling community.

Organization of the book

This book is organized into five chapters:

Following this introductory part, Chapter 1 presents the issue of whaling and its history. The circumstances under which the current international whaling regime was developed are described, followed by the outline of the stages of the regime's evolution. Following that, the history of whaling in Japan is discussed in detail. Japan's participation in the creation and development of the whaling regime is given particular attention.

Chapter 2 provides an overview of what has been written about whaling so far. Historical pieces of writing covering global whaling and whaling in Japan are discussed. This is followed by an introduction to the current trends in whaling research with a focus on the situation surrounding Japanese whaling.

Chapter 3 presents the conceptual and methodological framework that was guiding this project. Phenomenological and ethnographic methodological approaches are considered in the context of this project's research goal. Descriptions of data sources and data collection strategies are provided. This is followed by justifications for using in-depth interviews and fieldwork-based methods to answer the research questions posed.

Chapter 4 discusses how the chosen conceptual and methodological framework fits the issue in focus – Japanese whaling and the people involved in it. Particularities of conducting interviews and fieldwork in the context of the problem are outlined.

Chapter 5 presents the key findings of this research project, organized around what I call the pillars of modern-day Japanese whaling. I start by discussing the concept of a "super-whale" and its influence on the perception of Japanese whaling by the Western observers, as well as critically assessing the Western preoccupation with the question of "why" Japan still whales. The first part of the chapter explains why asking "why" might be premature in this case. Instead, we should be more curious and accepting of different perspectives through finding out more about the "who" and the "what" of Japanese whaling.

The first pillar is then presented, namely, the Japanese whaling community that upholds the traditions and pushes for innovations in how whaling is conducted and how whaling products are received by consumers.

Continuity is identified to be the second pillar of whaling in Japan. Although somewhat controversial – many argue that Japanese whaling lacks continuity since the traditional elements of it are largely lost, this part of the chapter argues that as long as continuity is what locals still believe in and strive for, it is not up to outsiders to decide on its absence. In several "whaling towns", as they are called by people in Japan, whaling has been an important part of everyday, both historically and into the now.

Han-han-hogei or anti-anti-whaling is identified as the third pillar the Japanese whaling is rooted in. The Western world's "anti-" has been critical in the development of the Japanese "pro-" in a process I call reactivity, borrowing this term from psychology. It remains to be seen how this might change now that Japan formally withdrew from the arena of the international whaling debates that is the IWC.

Orientation towards the future is the fourth pillar discussed in Chapter 5. This part is a refocus of attention from what has happened so far to what the Japanese whaling community aspires to achieve in the short- and long-term perspective.

Notes

1 See, for example, Takao (2012), who argues that Japan has fallen behind in global environmental innovation leadership.
2 On Kyoto Protocol (1997) see United Nations Climate Change information https://unfccc.int/kyoto_protocol
3 The word "cruel" in relation to Japan's treatment of certain whale species is used in many popular and academic publications, as well as during the International Whaling Commission Committee meetings by those who oppose whaling. The anti-whaling camp also prefers the word "kill" to, for instance, "culling" or "taking" – more neutral terms commonly used in wildlife management. Anti-whaling discourse, which includes the predominant Western media discourse, also does not highlight the fact that Japan targets only those species of whales that are proven beyond a reasonable doubt to be able to withstand a certain level of sustainable hunting. More examples and reflections on the overall censorious language used for Japanese whaling are presented in Chapter 5.
4 The countries members of the IWC had three years to accommodate the measure and the "pause" took effect from the 1985–1986 season onwards.
5 The concepts of "emic" and "etic" are from anthropological scholarly tradition. They will be discussed in detail throughout the book.

References

ABC (2009, December 11). Japan Vows to Carry on Whaling. Accessed online http://www.abc.net.au/news/stories/2009/12/11/2768448.htm?section=justin

BBC (2008, November 18). Japan approves whale meat import. Accessed online http://news.bbc.co.uk/2/hi/science/nature/7735355.stm

BBC (2015, November 28). Japan to resume whaling in Antarctic despite court ruling. Retrieved from http://www.bbc.com/news/world-asia-34952538

Chairman's Report of the Thirty-Fourth Annual Meeting (1982). Agenda Item 6. Accessed online https://iwc.int/document_3719

Danaher, M. (2002). Why Japan will not give up whaling. *Pacifica Review: Peace, Security & Global Change, 14*(2), 105–120.

Gales, N. J., Kasuya, T., Clapham, P. J., & Brownell, R. L. (2005). Japan's whaling plan under scrutiny. *Nature*, 435(7044). Accessed online http://www.nature.com/nature/journal/v435/n7044/full/435883a.html

Kolmaš, M. (2017). Japan and the Kyoto Protocol: Reconstructing proactive identity through environmental multilateralism. *The Pacific Review*, *30*(4), 462–477.

Lauer, T. (2020, April 23). Japan wants to catch whales. But who will eat them? Whalers depend on subsidies, as there is little demand for their catch. The Economist. Accessed online https://www.economist.com/asia/2020/04/23/japan-wants-to-catch-whales-but-who-will-eat-them

Lewis, S. (2019, June 29). Japan prepares to resume commercial whaling after 30 years, despite public outrage. *CBS News*. Accessed online https://www.cbsnews.com/news/japan-whaling-japan-prepares-to-resume-commercial-whaling-after-30-years-despite-public-outrage/

Lies, E. (2014, June 10). Japan shrugs off embarrassing court loss, vows resumption of Antarctic whaling. *Reuters*. Accessed online http://www.reuters.com/article/us-japan-whaling-idUSKBN0EL0YQ20140610

MOFA (2018). Global Issues and ODA: Environment. Ministry of Foreign Affairs website. Accessed online https://www.mofa.go.jp/policy/environment/index.html

Takao, Y. (2012). The transformation of Japan's environmental policy. *Environmental Politics*, *21*(5), 772–790.

Talmazan, Y. (2019). Japan resumes commercial whaling after three decades. NBC-News, July 1. Accessed online https://www.nbcnews.com/news/world/japan-resume-commercial-whaling-n1025046

Whiteman, H. (2015, December 1). Japan defies world as 'research' ship embarks on minke whale kill. CNN. Accessed online https://www.cnn.com/2015/11/30/asia/japan-whaling-research/index.html

1 Whaling dynamics throughout history

Whaling in only one country – Japan, is the main focus of this book. However, cetaceans are highly migratory species that travel thousands of miles, including in the high seas.[1] Japan, like most countries that have ever been involved in whaling, has a history of whaling in waters outside of its own marine territory, and thus being a part of the global whaling trends.

In the first decades of the 20th century, whaling emerged on the international agenda, and although the degree of attention to this issue has been fluctuating since then, it still concerns many nations across the globe now. Ever since whaling went global, Japan's participation in this industry has been of interest to the international community. How Japan has developed its relationship with whales and whaling has also been influenced by what people in other countries thought and did about it. In light of this interconnectedness,[2] it would be impossible to discuss the history of Japanese whaling without considering the background of how the global whaling regime started and evolved.

The organization and scope of this chapter reflects the importance of the events happening in the international arena as the context for the study of Japanese whaling and the people involved in it today. The first section is divided in two subsections and outlines how whaling and its management became a public and political concern globally. Japan's pursuit of the right to whale throughout its history is discussed in more detail in the second section. This also covers Japan's withdrawal from the International Convention for the Regulation of Whaling (ICRW) in 2019. Although the full effects of this decision are yet to be seen, it is clear that this is a historic development in this issue.

Some aspects of the history of whaling are also included in other chapters, as history is both the background and, as we will further see, an important factor in how people in Japan experience whaling and its products now.

The global whaling regime and its underlying norms

The issue of whaling has received a lot of publicity during the past several decades, which has largely overshadowed the centuries-long and complex history of this practice in the eyes of Western media consumers. To reflect and highlight this fact and to guide the reader from the known to the unknown, we proceed backward in

DOI: 10.4324/9781003255031-2

history – from the recent events and the attitudes toward whaling that most countries hold now to the deep roots of the modern impasse between the anti-whaling forces and pro-sustainable whalers.

Environmentalism and its impact on whaling – "preservation"

It is common to consider the 1972 United Nations Conference on Human Environment in Stockholm to be the birth of contemporary international environmentalism. Concerns about the consequences of human activities were, of course, voiced before. However, it was in the 1970s that environmental considerations became a lot more mainstream. The Stockholm Conference started the shift to a paradigm that emphasizes and, in some cases, even prioritizes the role of environmental protection in international relations (Abdel-Hadi, Tolba, & Soliman, 2010).

Whaling was one of the matters on the agenda of this first conference fully dedicated to environmental protection. Up to that point, during more or less just one decade – the 1960s, from an industry that was quietly accepted and only discussed among professionals, whaling became a key controversial matter regularly featured in popular and political debates. It is widely agreed that whaling played an important role in the rise of environmentalism in the 1970s. At the same time, environmentalism has also had a profound impact on the trajectory of whaling developments.

Starting from the 1960s, cetaceans have been gaining popularity in mass culture through books, television shows, recordings of the sounds that dolphins and whales make, and music inspired by the "songs" of whales,[3] as well as entertainment water parks featuring performances by these animals – all these convinced people that cetaceans were special animals with highly developed intellectual abilities (Ellis, 1999). These also helped construct a perceived bond between humans and cetaceans – focus on "perceived". From then on, dolphins and whales have been increasingly considered "our friends at sea", one of the most human-friendly species, and stories of them helping and saving seafarers spread, depicting the relationship between men and cetaceans as a mysterious and strong connection. This image of the "super-whale" crafted by media and popular culture, which incorporates qualities that in reality belong to different species of cetaceans, is given more attention in this book, in Chapter 5.

In parallel with the growing interest in dolphins and whales from the general public, a different "trend" was seen in the whaling world. By the late 1960s, the populations of several species of whales[4] had dramatically declined and were later added to the list of endangered species (IUCN, 2015[5]). Commercial whaling was believed to be the main reason for that (IUCN SSC, 2016), and this has become widely known to the public through publications in scientific journals and popular magazines (Ellis, 1999). The first anti-whaling protests followed shortly, and it can be said that these protests have been occurring ever since, with more or less frequency depending on whether whaling makes the news of the day. Thus, the 1970s marked the beginning of the whaling controversy, where the pro-whaling countries, including Japan, were facing strong opposition from the anti-whaling countries, groups, and individuals.

To most casual observers, the current whaling regime seems to be about the *preservation* of this species – a complete ban on hunting all species of whales based on the fact that they are all endangered (which is inaccurate) and supposedly possess superior qualities (which is debatable). It is accurate to view preservation as the currently dominant objective of the existing whaling regime. The institutional side of this objective is embodied in the so-called *moratorium on commercial whaling* – a "zero quota" on commercial whaling of all species that was introduced in 1982 and took effect in 1985 (IWC, n.d.). Although in many popular as well as academic sources, the moratorium is treated as a "ban" and "prohibition", and these words are used interchangeably with "moratorium", this in fact promotes an erroneous understanding of what this measure was originally designed to be. The moratorium was introduced as a pause – a decision to be reviewed in 10 years after more accurate scientific data would be gathered and presented to the interested parties. However, as a result of solidifying anti-whaling norms and perceptions in the majority of Western countries, this never happened – the provisions of the moratorium have never been reviewed, and it is still in full effect now.

"Management" and "conservation"

Looking at the evolution of the whaling regime backward, preceding preservation, there was the *management and conservation* stage. Already in the first decades of the 20th century, the whaling industry circles had concerns about the state of whale stocks. The 1931 Geneva Convention for the Regulation of Whaling marked the beginning of the global whaling regime in terms of international law. It was succeeded by the 1937 International Agreement for the Regulation of Whaling, and in 1946 by the *International Convention for the Regulation of Whaling*, which is seen as the foundation of the current global whaling regime. ICRW and the *Schedule to the International Convention for the Regulation of Whaling* – the part of the agreement that is revised on a regular basis – are the main legal instruments guiding whaling globally. As stated in ICRW, in the 1940s, the countries needed to work on "proper conservation and development of whale stocks" for "the orderly development of the whaling industry" (ICRW Preamble). Undoubtedly, back then, conservation of whales was seen as a means to an end – the desired outcome was the sustainability of the whaling industry rather than that of the species. Thus, the initial objective of the IWC was the proper *management* of whale stocks, which were then considered an important marine resource.

To achieve the goal of managing global whaling, ICRW provided for the creation of the *International Whaling Commission*. More than seven decades later, with 98 member countries[6] holding meetings biennially, this international organization is still a key pillar of the global whaling regime. From its inception to approximately the late 1960s, IWC was trying to fulfill the goal of managing whale stocks by issuing quotas for its members according to which they were supposed to regulate their hunts. However, the quotas were not limiting whaling enough, and some countries were later found to have been disregarding them completely (Ruffle, 2002).

Multiple stocks' depletion and the industry crisis followed, becoming the catalysts for the environmental movement and the norm of preservation discussed above. With this, we have come full circle in this short, backward historic overview.

Japan and the history of whaling

Since in this book Japan is the main player of interest on the global whaling scene, this country's relation to and with whaling history is given special attention. The prepositions here – "relation with and to" – might seem confusing, but there is indeed a relational complexity between the history of whaling and Japan. It is not just the history of whaling in Japan; neither is it just the history of whaling in general and Japan's part in it. History has played a significant role in the present-day whaling situation in Japan and globally in several different ways.

If we consider history to be linear, it is usually discussed as a precursor to whatever we witness at the moment, providing causal explanations and context. The following section on the history of whaling in Japan serves the same purpose, and this is no different from other research publications dedicated to long-standing issues.

However, in the particular case of whaling in Japan, history is more than just background information that should be investigated for a more profound understanding of the problem. In a way, the history of whaling in the world and particularly in Japan and how various stakeholders see it (or refuse to see it) and use it, is in itself a part of the problem. These two ways of looking at whaling history and Japan – the "*how it all started and unfolded*" and the "*argument in the debate*" – are both invoked in this book.

The "how it all started and unfolded" perspective on history fits into this chapter, which is mainly aimed at providing a starting point for the discussion and explaining the origins of the problem. The "argument in the debate" perspective, on the other hand, belongs in the chapter on findings of this research project, because only after some thorough desk research, as well as the fieldwork and the interviews, has it become apparent that history in this case is not only the background but also a part of the foreground – history was referred to by many of the interviewees, different aspects of it shaping their understanding of who they are now and what they represent as individuals and as a community.

Accordingly, the history that led Japan to where it stands on whaling is presented here below. The findings on whaling history as it is featured in the experiences of Japanese people involved in this practice are laid out in Chapter 5. These introductory paragraphs to both these parts are included here to stress that the two are closely intertwined. Their interrelation influences both the whaling issue in general and the people in Japan who engage in whaling, work on whaling-related policies, promote whaling, have whaling-related businesses, and/or consume whaling products. Accordingly, the sections that come hereafter are aimed not only at addressing the readers' curiosity about what was happening on the Japanese whaling scene over the past four centuries. They also hold clues to what will be discussed in the findings chapter.

Another important note must be made before the narration dives into pre-modern whaling in Japan. Numerous accounts of the history of whaling and specifically the history of whaling in Japan have already been given in other textual and visual pieces. It is impossible to reinvent history, and admittedly, there is a great deal of reiteration of what is already known here below. However, to add to the existing narrative, I took a slightly different approach in determining what events in the history of Japan should be highlighted in this book.

A list of major events in the history of whaling in Japan is presented on the website of the Japan Whaling Association (JWA)[7] – a Tokyo-based organization whose work I was observing during this project and whose employees were some of the people I had the opportunity to interview for it. JWA's offices are adjoined by the ones of the only major whaling company with capacity for pelagic whaling[8] left in the country – Kyodo Senpaku Kaisha Ltd., as well as with the Institute of Cetacean Research (ICR). This reflects the organizational closeness of the three and provides for their effective coordination and regular communication. JWA representatives also have projects in common with whaling-related governmental offices, both national and local, whaling-related small businesses and pro-sustainable whaling non-profits, such as Women's Forum for Fish, and individual activists.[9]

Given the interconnectedness of the different actors on the Japanese whaling scene (see Chapter 5 for further details), it is possible to suggest that the above-mentioned list of historic events reflects what JWA, but also other important actors in the Japanese whaling landscape, want to underline as the most notable developments in Japan's whaling history. A very similar list, titled "Partial Chronology of Whaling", is also featured on the website of Japan's Ministry of Foreign Affairs.[10] The MOFA's chronology ends with the year 1995, while the JWA's website has more recent entries – up to 2018 in the English version (the announcement of Japan's withdrawal from the ICRW) and up to July 2019 in the Japanese version (Japan resuming commercial whaling in its territorial waters and exclusive economic zone).

Assuming this list to be at least somewhat representative of what the Japanese pro-sustainable whaling circles believe to be important in the history of their whaling industry, instead of providing a generic historical overview aggregated from different neutral (or, rather, claiming neutrality) sources, I use the JWA-compiled list as a "skeleton" for the discussion of the history of whaling from the Japanese perspective. Using this list is comparable to borrowing the glasses through which the Japanese whaling-related people see the history of whaling, and this strategy fits well into the overall methodology and goal of this project – seeing "through the eyes of".

The JWA's chronology of whaling as found on the organization's website is presented in Table 1.1.[11] It does contain important global whaling developments. However, what is of more interest here is what was happening in Japan and which global events had the most influence on the developments in Japanese whaling. The following two sections will address the possible reasons why certain events are highlighted, as well as why certain language is used. Since the list provided by the JWA only features very brief notes for each of the entries, sources other than the

Table 1.1 Chronology of whaling (Japan Whaling Association's website, English version)

9th Century	Whaling starts in Norway, France, and Spain
12th Century	Hand-harpoon whaling starts in Japan
1606	Hand-harpooning whaling by organized groups starts in Taiji, Japan
1612	Hand-harpooning of Baird's beaked whales starts in Chiba Prefecture, Japan (near Wadaura)
1675	Whaling using nets begins in Taiji, and spreads to Shikoku and Kyushu, contributing to rapid expansion of whaling
1712	Sperm whaling starts in the United States (US-style whaling)
1838	Organized whaling using nets starts in Ayukawa, Japan
1864	Modern whaling is developed in Norway
1868	With harpoon guns completed in Norway, Norwegian-style whaling starts
1879	A storm claims the lives of 111 whalers from Taiji. This incident prompts transition from net whaling to modern whaling
1899	Japan starts Norwegian-style whaling
1903	The world's first whaling factory ship (Netherlands) sails out to Spitsbergen sea
1904	Norway sets up a whaling station in South Georgia Island; whaling begins in Antarctic Ocean
1905	First whaling factory ship sails to Antarctic Ocean
1906	Full-scale modern whaling starts in Japan with construction of modern whaling station in Ayukawa
1925	A mother ship equipped with a slipway goes on whaling for the first time
1931	First International Whaling Convention is signed
1932	Claws (tail fin pinchers) appears
1934	Japan enters mother ship-type whaling in Antarctic Ocean
1940	United States quits whaling
1941	Japan suspends mother ship-type whaling upon the outbreak of World War II
1946	International Convention for the Regulation of Whaling is signed Japan resumes whaling in Antarctic Ocean
1948	International Whaling Commission (IWC) is established
1949	1st IWC meetings are held
1951	Japan joins IWC
1959	Olympic system is abolished; Self-declared whaling starts
1962	Country quota system starts
1963	Hunting of humpback whales in Antarctic Ocean is banned United Kingdom quits whaling
1964	Hunting of blue whales in Antarctic Ocean is banned
1972	Resolution calling for 10-year moratorium on commercial whaling is adopted at United Nations Conference on Human Environment Blue Whale Unit system is abolished; catch quota by whale type system starts Norway withdraws from whaling in the Antarctic Ocean Japan starts minke whaling
1975	New Management Procedure (NMP) is adopted
1976	Hunting of fin whales in Antarctic Ocean is banned
1978	Hunting of sei whales in Antarctic Ocean is banned
1979	IWC adopts an Indian Ocean whale sanctuary
1982	IWC adopts a commercial whaling moratorium
1985	Japan withdraws objection to IWC moratorium

(Continued)

Table 1.1 (Continued)

1987	Japan withdraws from Antarctic whaling and starts research whaling (JARPA)
1988	Japan suspends coastal catching of minke and sperm whales
1990	IWC estimates population of minke whales in Antarctic Ocean as 760,000
1992	Iceland withdraws IWC; North Atlantic Marine Mammal Commission (NAMMCO) is established IWC completes development of Revised Management Procedure (RMP)
1993	Norway resumes commercial whaling
1994	IWC adopts southern ocean whale sanctuary
1994	Japan starts research whaling in northwest Pacific (JARPN)
2000	Japan starts research whaling (JARPNII) for feeding study
2002	54th IWC meeting is held in Shimonoseki
2003	IWC adopts Berlin initiative
2005	Japan starts research whaling (JARPAII)
2006	IWC adopts St. Kitts and Nevis Declaration
2007	Conference for the Normalization of the IWC in Tokyo
2012	IWC Scientific Committee agreed to 515,000 as a new estimate of the abundance of Antarctic minke whales
2014	ICJ judged to suspend the issuance of special permission for JARPA II implemented by Japan
	Japan submitted to the IWC Scientific Committee a new research plan in the Antarctic Ocean (NEWREP-A) replacing JARPA II
2017	Japan submitted to the IWC Scientific Committee a new research plan in the North-Eastern Pacific Ocean (NEWREP-NP) replacing JARPN II
	National Law on the implementation of scientific research on whales for the implementation of commercial whaling enforced in Japan
2018	Japan notified withdrawal from the International Convention for the Regulation of Whaling and announced resumption of commercial whaling starting July 2019

JWA were also extensively used to create a coherent and detailed picture of whaling in Japan throughout the centuries.

Whaling in Japan before the World War II

JWA estimates that hand-harpoon whaling in Japan started in the 12th century. The ICR – a Japanese government-backed organization that was in charge of the country's research whaling among other activities – in its brochure titled "Whales and Whaling"[12] gives the same date for the beginning of hand-harpoon whaling in Japan. However, in the same document, ICR also claims that targeting large species of cetaceans in Japan had already started during the Yayoi period – as early as the first-century BCE. Additionally, according to the ICR, the inhabitants of the Japanese islands were using products from stranded whales since the Jomon period – from about 7000 BCE. Takahashi, Kalland, Moeran, and Bestor (1989) state that although occasional whale hunting was practiced earlier, whaling was not established as a business in Japan until "well into the 16th century".

The timelines given by these three sources might seem disparate; however, they are not necessarily in conflict with each other – as with many aspects of the whaling debate (or any debate really), one needs to look beyond the general terms to have a clear picture. In this case, a lot depends on the distinction between "active whaling" and "passive whaling", on whether targeted cetaceans are small or large, whether hunting is done by a small or large and well-organized group of fishermen, as well as whether there are meat processing facilities established specifically for the purpose. Utilizing beached whales as well as targeting wounded ones drifting not far from the shore is considered passive whaling (Takahashi et al., 1989). Active whaling, as evident from its name, occurs when fishermen purposefully go out into the sea to catch healthy animals. In this type of whaling, some arrangements are also made in advance for processing the catch on shore. Most scholars, politicians, and commentators from various areas of expertise agree that the history of active organized whale hunting by large-scale enterprises in Japan is more than four centuries old.

In the early 17th century CE, organized group hand-harpoon whaling started in **Taiji**, Wakayama Prefecture. The town's geographical location and a lack of easy access to and from any major cities in the area traditionally made it almost entirely reliant on the sea as a means of survival (Endo, 2011). In 1606, a member of the then-powerful Wada family clan, Chubei Yoritomo Wada, established a so-called *kujira gumi* or "whale group", also called *sashite gumi* or "spearing group", in Taiji (Taiji Declaration on Traditional Whaling, 2007). *Kujira-gumi* is an important term in the history of whaling in Japan, as its origination signified the transition to a phase where whaling became a highly skilled, specialized, and structured economic activity.

In 1675, the third head of Taiji's whaling company, Kakuemon Yoriharu Wada, devised an original method of whaling – using nets first, followed by hand-harpooning (Takahashi et al., 1989). It has become known as *amitori ho* or "the net method". The nets were set up close to the shore, and dolphins or whales were driven into the nets from the ocean by groups of hunting boats. This hunting method was significantly more efficient than simply targeting animals in open water. After being successfully adopted in Taiji, it spread to other areas of Japan (Kita, 2007).

Amitori-ho has been particularly important for the establishment of whaling as the principal activity in Taiji because its organization and management involved many people with varied skills. During the prime years of whaling in Taiji, almost the entire population of the town was involved in one or another stage of whaling in some capacity. During the preparations for a whaling season, women in Taiji and the neighboring villages made hemp ropes; male experts in net weaving, or *ami-daiku*, made nets from the ropes – *ami-daiku* were sometimes invited from other provinces because of their highly specialized skills; boat builders, or *funa-daiku*, were also, of course, actively engaged, as well as those who made harpoons and other equipment (Takahashi et al., 1989). Whaling itself required the cooperation of several groups of people as well. Along with harpooners and those who operated the boats, there were also people who watched out for pods of cetaceans or individual whales out in the ocean from specific points on hilltops – *yamamikata* ("watchers from the mountain"). As soon as *yamamikata* noticed a group of dolphins, a whale or whales, they gave smoke signals to the crew on the shore. The latter then knew to take off in pursuit of the catch. This way of organizing the hunts makes the traditional Taiji style of whaling stand out – before departing in pursuit of their quarry with the whaling boats Taiji people waited for confirmation that the whales had come close to their shores, while in other whaling countries, the crews were going out into the sea just hoping they would find whales (Ellis, 1999).

There were three types of boats that participated in traditional whale hunts with nets in Taiji – *sekobune* ("chase boats"), *amibune* (net boats), and *mossobune* ("towing boats"), their names reflecting the role each type of boat had. Typically, around twenty boats in total went out into the sea at a time. All boats were made colorful and featured elaborate designs. Special attention was paid to *sekobune* that were made into veritable works of art with elements of nature and ornaments painted on them in great detail. Most of the original boats were lost at sea, but some of the designs were reconstructed from the remaining fragments and scrolls from those times.[13]

Typically, whalers were highly skilled in one part of the whaling process and would not shift to a different one; often, these skills were also transferred in the family from generation to generation. Each member of *kujira-gumi* was likely to have watched their elders do this work from an early age and was also likely to have started participating while young, which contributed to their mastery (Komatsu & Takagi, 2006). Such specialization helped to organize well-coordinated and highly efficient hunts.

Whale body processing on land was specialized work as well. Since in Japan whales were caught not only for blubber and oil, like in most European countries at that time, but also for food, the culinary requirements made flensing a task that required a high level of knowledge and mastery (Takahashi et al., 1989). Nowadays, flensers are rare specialists in Japan, and most of them originate from Taiji and other traditional whaling towns. There are 20–30 flensers now who travel around the country to lend their skills, and it is a tight group of people who all know each other well (personal observations and conversations held at a research whaling site in Kushiro, Hokkaido Prefecture, as well as in Wadaura, Chiba Prefecture).

Japanese people used various parts of a whale for food, which, depending on the species, included entrails, blubber, muscle, and certain parts of skin. This is especially remarkable because for many centuries killing animals and consumption of their meat were not encouraged in Japan in accordance with the postulates of Buddhism, which has been a prominent religion in the country since its import from China in the end of the 6th century – beginning of the 7th century. Whales, however, were considered big fish, which was and still is reflected in the Japanese language.

The Japanese character for "whale" is 鯨 (reads as *kujira*). Its radical,[14] which is the left component of the character in this case – 魚 – means "fish". The component on the right – 京 – is mostly known for denoting "capital", but it is a numerical as well – quadrillion.[15] In this latter sense, it can signify something "great, huge". An old Japanese word for whale is *isana*, written using two characters – 勇魚 – where the first one means "bravery" and the second one is the mentioned above "fish". *Isana* is still used for certain whaling-related events and names of organizations, for example, the Taiji Isana Festival or the Taiji Isana Association. We see that both the newer and the older Japanese words that denote a whale have the "fish" part to them. Another example of thinking of whales as fish is one of the words Japanese people use for flensing – 魚切 (reads *uo-kiri*), which literally means "fish cutting" (Takahashi et al., 1989).[16]

The assumption that whales were big fish to be treated like other marine resources made them an important exception to the Buddhist rules when Japanese people strongly adhered to them. Although some Buddhist priests were against it, whale meat gradually made its way to the most important tables of the country – those of emperors and shoguns, as part of the seafood menu (Japan Whaling Association, 1987). It became known as a delicacy and an important source of animal protein in these social circles. In parallel, people involved in whaling also received whale meat or other whale products as a part of payment for their work; they then used it themselves or exchanged it for other goods and services (Takahashi et al., 1989). As a result of these two processes, what is called *kujira shokubunka* ("whale food culture") spread and developed in Taiji and beyond.

Japan's *kujira shokubunka* is different from that of other countries. In Iceland and Norway – the two countries that currently still utilize whales as a food resource – primarily red whale meat is consumed (personal communication, August 2017). Other former whaling countries of the global West have never developed an appetite for whale meat, and their hunts were driven primarily by the demand for whale blubber and oil made from it, which was then used for energy production. Most whaling history researchers, even those who are critical toward Japan continuing the practice of whaling now, admit that traditional Japanese whaling was not wasteful – nearly the whole of a whale's body was used. Pre-20th century Japanese people would likely be shocked to find out that Europeans, Russians, and Americans were throwing away everything but the blubber.[17]

Other uses of whale products in Japan historically included boiling whale blubber and turning it into oil, soap, and pesticides that were highly valued by farmers; boiling whale bones to extract more oil and produce an effective fertilizer that was in demand across Japan; sinews and baleen were used for producing a varied range of items, including instrument strings, fishing rods, plates, fans' and umbrellas ribs; various internal organs were used to produce medicine, etc. (Ellis, 1999; Nicol, 1979b; Takahashi et al., 1989).

These days, encountering the word combination "brave whalers" in a modern English language text – spoken or written – is almost unthinkable. When people of this occupation are discussed, the adjectives used to describe them are usually far from flattering. The issue of the perception of whaling and whalers in the Western world is one of the central topics in this project and is revisited throughout the book. To give one example here, a reality TV show called "Whale Wars" aired on Animal Planet from November 2008 to January 2015. All six seasons (52 episodes) of the show consist entirely of scenes of how members of a marine advocacy and action group called Sea Shepherd Conservation Society chase the Japanese whaling fleet in the Southern Ocean. This organization's radical anti-whaling campaigning strategies, including physical and verbal harassment of the whalers, and the show's evident popularity (demonstrated through it being continuously funded and aired on a popular channel, as well as through relatively high ratings across online review platforms), reveal a general pattern of negative and even aggressive attitudes toward whalers.

The perception of whalers is very different in the communities who were historically fed by their labor. This might appear obvious, yet it is rarely meaningfully mentioned when the issue is discussed.

In the 21st century, whaling occurs in an extremely challenging environment and requires highly specialized skills. With what is called old-style whaling, it was even more dangerous and required more effort. In traditional whaling there was one whaler whose job was to climb onto the speared whale's back, cut out a hole in the animal's flesh, and tie the body to the ship with a rope put through that hole. A different person was in charge of diving into the ocean and fixing two large poles underneath the captured whale for these to support the animal's body and connect it to the two ships between which it was then towed to the shore (Takahashi et al., 1989). Moreover, both of these daunting parts of a whaling operation had to be carried out while the whale was still alive to prevent the body from sinking. People who live in places where whaling is being looked at favorably think of whalers in more positive terms and see them as courageous, fearless, and strong, as everyone is aware of the risks associated with whaling and the concurrent physical and mental stress of working at sea for prolonged periods of time. This was especially true in the past, in towns where whalers were the strongest providers for the whole community, and whaling in one way or another was employing many of the community's members. When whalers are the main breadwinners (or rather protein winners) – they are well respected and looked up to.

In this context of support and appreciation of whalers and their work, as well as while most of the community was supported by whaling, a grave whaling accident occurred in Taiji at the end of December 1878 (marked as 1879 in JWA's list of historical events). The year 1878 was one of the least successful for the whalers of Taiji – they did not manage to catch a single whale until the very end of it. This was possibly caused by the fact that American whalers had by then discovered the abundance of right and sperm whales in the waters off Japan and were engaged in massive hunting operations in the western North Pacific (Kita, 2007). Although the people of Taiji were not aware of that back then, the Americans presented strong competition to the Japanese whalers. That was because, as discussed earlier, in Japanese traditional whaling, they waited for the whales to come close to the shore rather than going out to the ocean to search for the catch. That year, not many of the targeted species reached the area where they could be spotted from Taiji's lookout posts.

On December 24, 1978, a right whale with a calf were seen swimming off the town's shore. It was not allowed in Taiji to pursue a female right whale if it was with a calf. As a saying from those times goes, "Even in a dream, look not upon a right whale and her calf" (Nicol, 1979b). Whalers were a superstitious group and took the saying very seriously, believing that grief would befall those who challenged it (Komatsu & Takagi, 2006). It should be said, though, that this warning was based on facts rather than a myth. People who were working with whales their whole lives likely had developed a good intuitive grasp of the principles of sustainability and were not intent on taking mothers before their young were fully independent. Additionally, approaching a female with a young is known to be a dangerous undertaking, as in most cases they would respond aggressively, protecting the offspring.

In Taiji, there was a special sign for cases of viewing a female right whale and a calf – raising three black pennants each with one white vertical stripe in the middle (Ellis, 1999). For whalers waiting at the chaser boats, seeing these pennants meant that there were these two whales out in the water, but that the hunters ought to remain on shore. Yet that afternoon, desperate to secure food before the New Year's Eve, the people in charge of the whaling operations decided to have the boats pursue the quarry. They managed to reach the whales, and the adult one was entangled and harpooned. But it was extremely big and strong, fought, and ended up dragging the boats with the people onboard far out into the ocean. The prolonged battle with the animal lasted into the fall of night, and the boats got separated from one another in the chaos and darkness. The next morning, a heavy storm started and lasted several days. The risky endeavor turned into a tragedy – the lives of most of the town's breadwinners, along with all of their equipment were lost. Only a few members of the crew were saved. Sources disagree on the exact number – but between 110 and 130 people died during these few days. This incident is known as *O Semi Nagare* (大背美流れ) or literally "the great right whale drifting away", and is considered the darkest day of whaling in Taiji and even the whole history of whaling in Japan (Komatsu & Takagi, 2006). After *O Semi Nagare*, it was extremely difficult for the town to organize a crew and rebuild the fleet for both economical and emotional reasons. In fact, these tragic events are thought to have brought the traditional style of whaling in Taiji to an end (Endo, 2011). After this, Taiji turned to the so-called modern style of whaling.

Taiji has retained its status as an important whaling center in Japan for centuries, and it still occupies a very important space in the landscape of Japanese whaling today. It is not only a geographical point on the map of Japan where this practice has been pursued for four centuries now. As Taiji spearheaded innovations in this domain and gradually developed a narrow specialization as a whaling town, it also gained considerable cultural and political significance in the context of the issue of whaling, mostly nationally but recently also internationally.[18]

Another location mentioned in the JWA's list of whaling-related historic events in Japan is **Wadaura**. Wadaura lies on the Pacific coast of the Boso Peninsula, and it is currently a part of Minamiboso city in Chiba Prefecture, around 70 kilometers south of Tokyo. The place has been known for providing nearby Tokyo with seafood since the early 17th century. As JWA states, hand-harpoon whaling started as a practice not far from here in 1612. To be precise, in 1612 it was organized in Katsuyama on the other side of the Boso Peninsula, then the whaling station was moved to Tateyama, after that to Shirahama, later to Chikura, and only then to

Wadaura (Kalland & Moeran, 1992). The JWA still indicates closeness to Wadaura with the 1612 entry instead of any of the other locations, as it is in Wadaura that whaling plays an important role in the life of the community in present-day Japan.

Wadaura's own whaling history is not as long as that of Taiji – the first whaling company called Gaibo Hogei was established here in 1948. The enterprise is still in operation under the leadership of its second CEO Yoshinori Shoji, who is the son of the founder. Gaibo Hogei historically focused on catching small species of whales – in Japanese professional whaling terminology, this type of whaling is called small-type coastal whaling (STCW) and is often abbreviated to STCW.[19] This specialization allowed the company to stay in business after the moratorium on commercial whaling was introduced in the 1985–1986 season. Since the ICRW and subsequently the IWC, as well as the moratorium, only regulate large cetaceans, the treatment of small species of whales is left for the national and local governments to control. Until 2019, when Japan restarted commercial whaling, Gaibo Hogei was receiving its annual quotas of 26–35 Baird's beaked whales (a "small type" by both Japanese standards and the IWC's) from the Ministry of Agriculture, Forestry and Fisheries of Japan, but was not allowed to catch minke whales it also formerly targeted – on the grounds of this species falling under the jurisdiction of the IWC as one of the 13 great whales. There used to be another whaling company partially operating in Wadaura that focused on the so-called large-type coastal whaling, or LSTW – Nitto Hogei, but it closed its local whaling station in 1987 after the commercial whaling moratorium came into effect (Kalland & Moeran, 1992).

Even though factually modern-day Wadaura is a relatively new whaling community, the general area of the Boso Peninsula has been involved in whaling for over four centuries, and the people here are used to having whale meat in their diets. The owner of Gaibo Hogei Shoji considers the long history of this practice a reason for him to not only to see his company as a business but also take a proactive stand on the issue of whaling:

> It [whale meat] is a resource like any other. We have had a chance to eat whale meat for 400 years ... and that is why I intend to work for the continuation of this [tradition] with respect.
>
> (personal communication, December 2016)

Wadaura's whalers consider their craft and its results not only a part of the town's history, but also its future, and they support Japan's withdrawal from the international treaty that restricted whaling. After the resumption of commercial whaling in Japan in summer 2019, it was a boat from Minamiboso that took the first minke whale,[20] and the locals took pride in that fact (personal communication, March 2023).

Another important location for Japanese whaling is **Ayukawa**. It is a small port in Miyagi Prefecture located on the edge of the Oshika Peninsula that juts into the Pacific Ocean from Sendai – the capital of Miyagi. Whaling was spreading from Taiji up toward the north, and it took at least a century and a half for it to reach this location in northeastern Japan. Ayukawa is included in the JWA's chronology twice. According to it, organized net-method whaling started here in 1838, and modern whaling in 1906. Kalland and Moeran (1992) provide only one year for the advent of whaling in Ayukawa – 1906. This divergence can be explained by the fact that what started as whaling using nets in 1838 was a short-lived attempt at developing this industry. It proved not very lucrative – according to some accounts, it was because of the already existing scarcity of humpback and right whales in the area due to overhunting by foreign parties (Nicol, 1979a). The operations were halted after about 5 years, with a total take of 40 whales (Ibid.). And it was in 1906 that the modern, or Norwegian-style, whaling came to Ayukawa, bringing considerable changes to the town's life.

When Japan turned to modern whaling, there were 12 companies engaged in this industry operating around the country. Due to the convergence of the warm Kuroshio and cold Oyashio currents, the Sanriku-oki area[21] was (and still is) rich in marine resources. In the beginning of the 20th century, these fishing grounds attracted 9 of the 12 whaling companies to Sanriku Coast, particularly Ayukawa, which became a bustling center for the development of whaling (Ishinomaki City, 2006). In turn, this influenced the demographic of this previously quiet fishing village. Whalers from older whaling communities moved here and brought their skills with them to pass on to the younger generations (Japan Small-Type Whaling Association, 2002). At the turn of the century, the population of Ayukawa was below 500 inhabitants, then right after a modern whaling station was built here in the early 1900s the town started to grow, and this trend continued along with the development of whaling throughout the next five decades – in 1955, 3795 people lived here (Kalland & Moeran, 1992). In the post-war years, when the Japanese whaling industry experienced a boom due to the scarcity of other sources of animal protein, Ayukawa's population grew even more, reaching its record population of about 10,000 and becoming increasingly busy (Fackler, 2011). In the 1990s, however, the population equaled approximately 5,000 inhabitants, with more than 50% of them over 50 years old, which was evidence of decreasing economic activity and opportunities as less and less whaling activities were carried out (Ibid.). The IWC-imposed moratorium on commercial whaling played the biggest role in what some considered "the danger of extinction" of not only the local whaling traditions, but of Ayukawa itself (Heazle, 1993). Minke whale was the main target for local whalers, and its management falls under the IWC jurisdiction. Although this species of whale has never been considered endangered, its catches were banned along with other great whale species all over the world, including Ayukawa and other Japanese whaling towns that used to hunt it. Since the moratorium has never been reviewed and lifted, in the early 2000s, Ayukawa continued to experience considerable population decline and an aging trend, according to the Japan Small-Type Whaling Association (2002). In 2010, the town's population size was 1,462 residents according to the national census (Tani, 2012). In 2011, the Tohoku earthquake and tsunami almost destroyed the town of Ayukawa, including its whaling facilities and equipment. To this day, the town and its remaining inhabitants are struggling to recover from this tragedy.

Regardless of its decline, Ayukawa is still considered a *"kujira no machi"* - town of whales.[22] Along with Taiji and Wadaura, it is one of the few locations in Japan that received quotas from Japan's Ministry of Agriculture, Forestry and Fisheries for taking small species of whales under STCW Regulations while Japan was a member of the IWC.[23] Wadaura and Ayukawa have also been connected through the whaling company that was introduced above – Gaibo Hogei Ltd. Its founder and his son – the current owner of the company, Yoshinori Shoji, hail from Wadaura, and that's where Gaibo Hogei's head office is located. However, in 1969, they established a branch in Ayukawa, engaging in STCW as well as making and distributing whale meat products in this northern location. As with the rest of the town, the company suffered great losses in the 2011 Tohoku earthquake and tsunami.

Ayukawa is also one of the ports from which the coastal component of the northwestern Pacific part of Japanese government research whaling – Japanese Research Whaling Program in the North Pacific, second phase (JARPN II) – was conducted. The program itself started with offshore survey feasibility studies in 2000 and 2001, followed by actual surveys from 2002. The coastal component of it – using small-type whaling catchers in the area of up to 50 miles off the coast – was initiated in 2002. In 2003, the boats departed specifically from the Ayukawa port for the first time.

The above discussion of the whaling history in Taiji and Ayukawa featured a reference to so-called modern whaling. The division between pre-modern, traditional, or old-type, as the Japanese sources often call it, and modern whaling is valid not only for Taiji and Ayukawa, and not even just Japan, but the whole (previously) whaling world. Modern whaling originated in Norway in 1864, as the JWA's website duly notes in its chronology. From there, it quickly spread all over the whaling world. It was a shift of great magnitude in whaling techniques that allowed to target faster species of whales and contributed to the immense growth of commercial whaling. The right whale[24] was one of the most sought-after species around the world during the pre-modern whaling era. Its stocks were in critical condition toward the end of the 19th century, and only by developing the capacity to target other baleen whales could the industry be saved from becoming unprofitable. This is exactly what happened. Steamboats and later diesel-driven boats were introduced instead of rowing boats; a harpoon was fired from a cannon instead of being manually thrown. Moreover, harpoons were now also equipped with a grenade, which exploded inside a whale's body after it was speared, and a line that allowed whalers to pull the quarry out of the water when it sank after being fatally hit (Tønnessen & Johnsen, 1982). Another Norwegian innovation was a technology of inflating the bodies of large baleen whales to prevent them from sinking after death, which enabled them to easily pull the catch up to the ships (Komatsu & Takagi, 2006). With these technological advances, modern hunts became incomparable to old-style whaling – both in scale and, subsequently, in the damage to whale stocks worldwide.

It took more than 30 years for Japan to modernize its whaling techniques, and when it did, it was certainly in response to other nations already actively utilizing more advanced ships and equipment and taking more whales in shorter amounts of time, including in the waters that used to feed Japan. The traditional way of waiting for whales to appear in sight that Japan practiced during the previous centuries was becoming extremely inefficient as the stocks were being quickly decimated by American, French, English, and Russian whalers. In the absence of international regulations and delimitation of maritime boundaries, Japan was left with no choice but to follow that global trend if it wanted to continue claiming a part of the marine resource in question. The first modern whaling company was established in Japan in 1899 (see the JWA's chronology). The start of this type of whaling in Japan is associated with the name of Juro Oka, who was a businessman and envisioned Japan becoming a great whaling nation. He traveled the world to learn whaling techniques and ended up importing equipment from Norway, hiring some Norwegian gunners, and starting a Norway-style whaling company, Nippon Enyo Gyogyo Kaisha (Komatsu & Takagi, 2006). Oka's company remained strong on the national whale meat market, but soon there were a few more modern whaling companies operating around Japan.

Winning the Russo-Japanese War of 1904–1905 brought several Russian factory-ships and catcher boats into Japanese possession, as well as the Nikolai warship, that was also repurposed into a whaling vessel. That substantially expanded the Japanese whaling operations – the two biggest companies on the market increased their capacities threefold, new whaling grounds were opened, new whaling companies appeared on the market, as well as land stations (Tønnessen & Johnsen, 1982). Later on, however, as the competition over the whaling grounds off the Japanese coast grew more intense, several mergers also occurred (Ellis, 1999). Throughout the 20th century, there was a tendency for mergers in the whaling industry of Japan. The only big whaling company with the capacity for pelagic whaling (see the explanation for this type of whaling below) that is still on the market in 2023 – Kyodo Senpaku – was also founded as a result of a big merger in 1976.

One of the most important developments in global whaling was the start of exploration of the Antarctic whaling grounds. It began in 1904 with operations set up in the southern Atlantic Ocean from the island of South Georgia by Norwegian whalers and developed gradually, with a few more Norwegian and British companies coming into the area. For about two decades, the hunts were centered around land stations that were being built on South Georgia, South Shetland Islands, and the Antarctic Peninsula. The licenses for whaling operations were administered by the British government's Colonial Office. That was legally based on the British sovereignty over the Antarctic harbors and, as the international law of the seas back in the day stipulated, the 3-nautical-mile territorial waters around those.[25] Britian was already starting to realize the possibility of overexploitation of whales as a resource and was making efforts to prevent that from happening through a number of instruments, including limiting the licenses, limiting the catches, and discouraging waste (Ellis, 1999).

But then the industry took a leap in technological advancement one more time and things changed again, for the better for whalers – but for the worse for whales. The JWA's chronology acknowledges the importance of this shift, as it has an entry for the year 1925, when a mother ship with a slipway was first tested out in practice (Table 1.1). Attaching a stern slipway to a ship meant that whales' bodies could be brought onto the ship and processed right there instead of dragging them to land for flensing (Ellis, 1999). That, along with some other innovations, such as introducing devices for the production of fresh water on board, essentially meant that the whaling industry gradually gained the capacity of having all parts of its operations carried out in the open sea without having to go into the harbors in the Antarctic. A new phase began, giving birth to a new category of whaling that is referred to as "**pelagic whaling**". Since the necessity to rely on harbors and land stations was dramatically reduced, whaling permits from Britain became legally ineffective.

Pelagic whaling between 1925 and 1965 was the most lucrative marine-resources-based industry in terms of the weight of the catches as well as its monetary value (Clark & Lamberson, 1982). That unparalleled effectiveness was what eventually brought both whales' populations and the industry feeding on them to the brink of extinction.

Japan joined the two other pelagic whaling nations operating in the Antarctic – Britain and Norway – in 1934 (see the JWA's chronology). As mentioned above, it was encouraged by the noticeably decreasing supply of whales in its own waters and the success of the competitors in the field. Japan bought a factory mother ship Antarctic (renamed to Tonan Maru) and hired Norwegian whalers to be onboard supervisors (Takahashi et al., 1989), but the core of the crew were men from Taiji (Nicol, 1979b). In a mere few years, Japanese pelagic whaling expanded so much that it became more economically important than whaling in Japanese waters. However, both continued to co-exist. For the next five decades or so, the Japanese whaling complex consisted of STCW, LTCW, and pelagic whaling. Japan continued to grow its whaling capacity, acquiring ships and building its own, and by the end of the 1930s, it became one of the most active whaling nations in the world (Ellis, 1999).

It was the start of World War II that curtailed the expansion of the industry. During the 1940–1941 season, there were as many as six Japanese whaling fleets out in the Antarctic Ocean. However, starting in the following year, all whaling in this area was suspended. Japan repurposed its whaling ships for military use, and most of the whaling fleet personnel was engaged in the navy. Some whaling still occurred during the war to provide food and oil, but the operations were limited to Japanese waters (Ellis, 1999). By the end of World War II, all the Japanese factory mother ships were destroyed, and many catcher boats suffered the harsh consequences of military campaigns (Komatsu & Takagi, 2006).

The next phase of whaling in general and Japanese whaling in particular started after the end of World War II. This time in whaling-related history is often paid a lot of attention to in both Japanese and English-language sources, as it is thought to have determined the course of the pro- and anti-whaling debate to this day. Two parallel tendencies were gradually developing, each having a significant impact on both the actual state of affairs in whaling and people's perceptions of it. First, Japan was hungry, and it was whale meat that was destined to alleviate that postwar symptom. Whale meat became the main source of animal protein, and the role of the whaling industry was on the rise in Japan. On the other hand, the rest of the world was getting increasingly anxious about whale conservation in the context of collapsing whale stocks. The whaling industry outside of Japan was also on the brink of a collapse. It was around that time that the debate between the pro-sustainable whaling position and the anti-whaling position started to take on its current shape.

International Regulation of Whaling and Japan's participation in it

Compared to the long history of people utilizing whales and whale products, the history of people trying to regulate this activity on the international level is relatively short. Japan was not among the first to get involved in the global whaling regime. It also ended up withdrawing from the leading international legal instrument regulating global whaling – the ICRW – in 2019. Regardless of these two facts, Japan was and still is one of the most vocal nations when it comes to discussing whaling policies. Professional observers noted that in no other international forum had they seen the representatives of Japan being as pro-active and assertive as they were in the framework of the IWC activities (personal communication during the IWC67, September 2018). Japan became known for promoting pro-sustainable whaling goals and values, and no efforts made by anti-whaling nations, organizations, and individuals managed to stir Japan off its course.

The principal international-level agreement governing whaling is the ICRW. As mentioned earlier, it was signed in 1946. This piece of legislation is still in force today and is regularly brought up during whaling-related fora and in the context of anti- and pro-whaling debates. And it is from the ICRW that Japan withdrew on July 1, 2019, which also meant the country was no longer a member of the IWC – the organization established by the ICRW.

Some earlier attempts to introduce restrictions and rules to the industry were made in the 1910s, 1920s, and 1930s. In 1913, the International Commission on Wildlife Protection expressed concerns about the poor state of certain cetacean stocks, and international action to control whaling was first proposed in 1918 (van Drimmelen, 1991). In 1925, the League of Nations officially acknowledged the issue of overharvesting of whales (Ellis, 1999). That happened thanks to the work of an Argentinian scholar, Jose Leon Suarez, who at the time served as the organization's Committee of International Law reporter. He wrote a memorandum on the matter, stating that the whaling industry was threatening some whale species' survival, especially in Antarctica, and that it was the League of Nations that had to advise the participating governments on possible ways of preventing further damage (Christol, Schmidhauser, & Totten, 1976). In 1927, the United States suggested summoning an international conference to discuss whales' conservation, but this proposal was not well received by other nations. In particular, Japan countered this with an assertion that the issue was better dealt with bilaterally or multilaterally between parties directly involved (Christol et al., 1976). Norway – one of the few currently whaling nations – was one of the first to initiate whaling regulations on a national level. In 1929, the Norwegian Whaling Act was drafted – a document, which inspired and informed the first ICRW under the auspices of the League of Nations (Ellis, 1991). This preliminary ICRW is also known as the Geneva Convention and was drafted and signed in this Swiss city in 1931. It took 4 years for this agreement to come into force, as it was only in 1935 that the number of ratifications reached the necessary minimum (Gambell, 1993). This and the next convention on the regulation of whaling – the Whaling Convention of 1935 – were largely the initiative of Great Britain and the USA, where a conservation movement started getting momentum around that time. The USA also already had national whaling legislation in place by that time – the Whaling Treaty Act of 1932 (Christol et al., 1976).

Although these early international whaling-related agreements could be considered a positive development, they were not successful in attaining their main goal of slowing down the decimation of vulnerable whale populations. They were not comprehensive and not as efficient as was needed already then. That was largely due to the fact that some of the key players on the global whaling scene at that time were not interested in whale conservation. Germany, the USSR, and Japan did not sign the Convention of 1931. As noted in the previous section, Japan was comparatively late to start operations in the Antarctic. With the waters close to its own territory depleted and having just gained access to the rich whaling grounds of Antarctica, the country was unwilling to compromise its immediate profits (Nagtzaam, 2009).

The next attempt at reaching an accord on the international level was made in 1937. An international conference on whaling was held in London with eight participating countries. Britain and the USA were joined by Argentina, Australia, Germany, the Irish Free State, New Zealand, and South Africa. A document known as the London Agreement, or the International Agreement for the Regulation of Whaling, was signed as a result of this meeting, and together with a Protocol to the Convention introduced in 1938, it is considered to be the first comprehensive whaling-related agreement (Nagtzaam, 2009). It was an important step forward in international deliberations on ways of controlling the whaling industry, as it covered operations in the sea as well as on land, defined whaling seasons by areas, and prohibited taking whales from certain stocks. However, the same problem as before was hindering the effectiveness of this agreement – some of the states actively whaling at that time, including Argentina, Chile, Germany, Japan, and the Soviet Union, were not interested in being bound by the treaty's provisions (Ellis, 1999). Along with the absence of enforcement and a poor scientific base, this resulted in the total figures showing not less but more animals hunted around the world and more oil harvested during the whaling season following the convention. And that is even though the countries' reports on the numbers of whales taken were not always accurate back then (Nagtzaam, 2009).

In this context of failing regulatory mechanisms and virtually uncontrolled taking of whales, it was World War II that slowed down some of the most active whaling nations and saved some species of cetaceans from extinction (Gambell, 1993; Nagtzaam, 2009). As was mentioned in the previous section, Japan's fleet was almost entirely deployed for military use, and most of it was subsequently damaged or lost completely.

To summarize, the early attempts to regulate whaling on the international level were not successful, as the nations that had the most knowledge of the problem and the potential to stop its aggravation were also the ones with the most developed and profitable whaling industries and had no desire to curb their own whaling activities. Later on, when certain regulations were agreed upon, the main motivating factor was not the health of whale stocks per se but the concern of the whaling nations about the industry's dropping revenues caused by over-harvesting and the abundance of whale oil on the markets, and hence its plummeting price.[26] It was the convergence of the interests of the whaling industry, which finally saw its own death in the uncontrolled hunts of whales, and the groups concerned about the environmental impact of over-whaling that brought about further, more advanced, developments in the regulation of whaling.

A standard measure for the amount of harvested oil – the so-called Blue Whale Unit or BWU – was conceptualized in the 1944 Protocol, which was, again, adopted at the initiative of Britain. The goal of having this measure was to limit the catches, regardless of the species the oil was coming from. BWU made one blue whale equal to two fin whales, two and a half humpbacks, and six sei whales in terms of the oil produced from each animal. BWU is another example of how international management initiatives in whaling turned out to be mismanagement in the end. The intention was seemingly sound in those circumstances – to have the hunts under control, as the limit was initially set to 16000 BWU for the 1945–1946 whaling season, which was about two-thirds of the average pre-war amount (Nagtzaam, 2009). However, the BWU-based catch limit system was later severely criticized as lacking a scientific basis, being fixed (it was not adjusted often enough to reflect the current state of stocks), and consequently contributing to the decimation of all whale species and populations with no regard to their actual status. BWU was actively in use until 1972, which is a clear indication that whale oil production was prioritized over whale conservation up until that time (Morishita & Goodman, 2005).

The JWA's chronology of whaling mentions only the Geneva Convention of 1931 before jumping to the International Convention on the Regulation of Whaling of 1946. This could be associated with how little Japan participated in these early regulatory discussions and arrangements.

The International Convention for the regulation of whaling and the International Whaling Commission

After the end of World War II, Europe was experiencing a shortage of edible oil, among others, and Japan was famished overall. This made whaling nations interested in reviving the whaling industry (Komatsu & Takagi, 2006). However, the same issues in international whaling management pertained. In this context, another conference was gathered in Washington, D.C., in 1946. The initiative was led by the USA, although they had nearly given up commercial whaling by that time.[27] A new ICRW was signed, establishing the IWC. At that time, there were 15 countries-signatories of the ICRW, and those also became the first 15 members of the IWC. This agreement went into effect in 1948.

A key part of the ICRW is a legally binding Schedule, which is a collection of operative regulations open to amendment by voting of the IWC members. A three-quarters majority is required to pass any such amendment. The Schedule allows for flexibility and adjustment of the regulations to reflect any newly obtained scientific data and, as time has shown, also new norms and perceptions of whales and whaling. A member-state can formally object to any changes to the Schedule within 90 days after they were voted for, which gives an opportunity for countries to avoid being bound by amendments they consider unfavorable to their own national interests (Gambell, 1993).

There are six committees within the IWC at present (see the IWC organizational chart below). The one that has been getting, perhaps, the most attention in the media and academic literature on the issue of whaling is the Scientific Committee (SC). It was established by the IWC in 1950, and its main tasks include recommending and conducting whale and whaling-related studies, collecting and analyzing data on whale stocks, as well as reviewing the IWC member-states' whales-related research programs, and commenting on permits issued for scientific research (Scientific Committee Handbook, 2018). Both pro-sustainable whaling and anti-whaling nations, organizations, and individuals claim to use science to support their policies and arguments, which makes the work of the SC extremely significant in the context of the pro- and anti- whaling debate.

Japan was not among the 15 initial IWC member states. This was due to the fact that, after its defeat in World War II, it was under the Allied occupation and could not sign any international treaties during 1945–1952. However, starting from the 1948–1949 whaling season, Japan was already sending its ships to the Antarctic, which was authorized by General Douglas A. MacArthur, who oversaw the Supreme Command of the Allied Powers in Japan. This was motivated by two

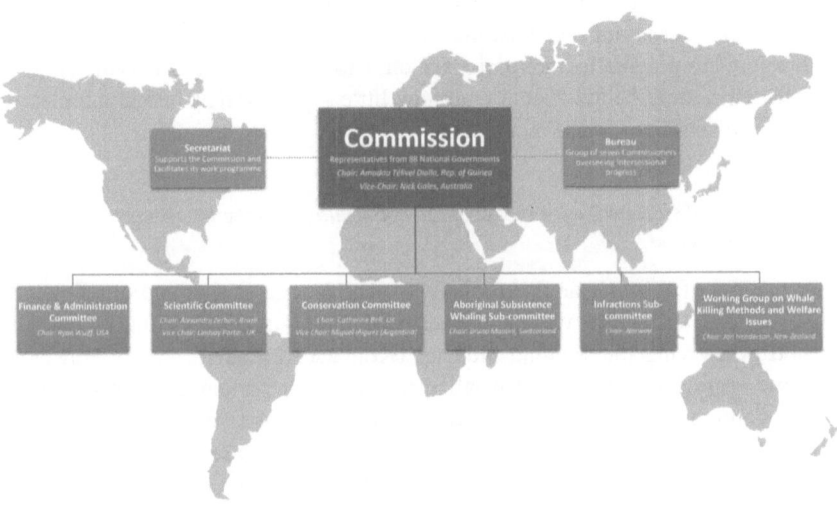

Figure 1.1 The IWC Organizational Chart (IWC, 2022).

factors: the need to feed the starving Japanese population and the US interest in getting whale oil. In 1951, Japan formally adhered to the ICRW and became a member of the IWC.

The purpose of the ICRW is "to provide for the proper conservation of whale stocks and thus make possible the orderly development of the whaling industry", as stated in the document's Preamble (ICRW, 1946). This is arguably one of the most quoted sentences in the pro- and anti-whaling debates, both oral and written. Once again, it is evident that at the time when the ICRW was drafted, the concerns for the whales' survival were secondary to the concerns about the whaling industry's survival – although the Preamble also features mentions of whale stocks as "great natural resources" worthy of "safeguarding for the future generations" (ICRW, 1946). To the modern-day animal rights' activists this order of priorities is unacceptable, as is "developing" the whaling industry. Representatives of the anti-whaling views would usually stress only the first part of the above Preamble statement – "for the proper conservation of whale stocks". However, for Japan and its allies the full goal of the ICRW and the IWC as originally formulated still reflects their understanding of the problem. They view whale stocks as a valuable natural resource, subject to science-based management and utilization. As some pro-sustainable whaling supporters, including some of those interviewed for this book, would sometimes say – whales are "just like any other marine resource" (personal communication and observations, 2016–2019). But it is important to note that on the scientific and policy levels, all factors that make the various cetacean species *not* like any other marine resource are taken into careful consideration when sustainable catch limits and management procedures are being devised by specialized institutions and professionals.

Although on paper the regulations presented in the new ICRW looked robust, covering the hunting seasons, hunting methods, catch quotas for various species, and other areas of concern, in reality, these measures did not improve the state of whale stocks. On the contrary, in the first three decades of the IWC's existence the numbers decreased dramatically for some species, and some came close to extinction. Several factors were at play here. Firstly, the IWC was what some called a "whalers club" – most members were whaling nations and had a vested interest in keeping the catches' limits high (Skodvin & Andresen, 2003). For that reason, the instrument of Objection to Schedule amendments was used too freely by the member-states, which made the necessary changes almost impossible to introduce (Nagtzaam, 2009). Additionally, not all whaling nations joined the IWC at that time, which was problematic as they were not bound by the IWC regulations and proceeded with unchecked catches. For instance, Chile and Peru adhered to the ICRW later, in 1979 (Ellis, 1999). The IWC Scientific Committee (SC) recommended lower catch quotas in the 1950s; however, those were not accepted. Additionally, no specialists in the then-emerging field of population dynamics were engaged in the work of the SC back then, which made it easy for the interested parties to discredit the committee's assessments (Morishita & Goodman, 2005).

As was discussed earlier, the late 1960s to early 1970s was a period when the environmental protection movement in the West gained in popularity. The poor state of cetacean stocks was a big influence on the emergence of an environmentally conscious worldview, and the opposite was also true – the spreading environmentally conscious worldview greatly influenced the management of cetaceans worldwide. With considerable pressure from eco-activists, a resolution on introducing a ten-year moratorium on commercial whaling was adopted in Stockholm at the United Nations Conference on Human Environment in 1972 (Morishita & Goodman, 2005). Politically, this reflected the mood of the times. However, the IWC SC reviewed the proposal that same year and recommended that there was no scientific justification for introducing a moratorium on hunts of all great whale species – what is often referred to as a blanket moratorium – according to the participating scientists, some stocks were proven to be abundant enough for sustainable hunts.

As in the Western countries, the concern about the environment in general and whales' lives in particular grew during the three post-war decades, what grew in Japan was the appetite for whale meat. The whole time before WWII, whale dishes were mostly a regional delicacy; after the war, they became popular all across the country. According to the testimonies of people who still remember those days, for them, "*niku*" ("meat") was the same as "*kujira*" ("whale" or "whale meat" in this context) – it was the only meat they knew (personal communication, numerous interview participants, and conversations, 2016–2019). The tradition of utilizing all parts of a whale for food also continued into this post-war period and into the present.

The two contradictory trends – of the West developing feelings for whales and Japan (further) developing a taste for whale meat – found their reflection in the IWC deliberations. Japan was generally not supportive of cautious catch quotas, and even less so of a moratorium. In 1972, when the idea of a ten-year pause on commercial whaling was first expressed during the United Nations Environmental Conference, Japan was among the six countries who opposed it at the IWC, along with Norway, Iceland, Panama, South Africa, and the Soviet Union (Day, 1992). Popular protests followed these countries' rejection of the moratorium, which attracted even more attention to the issue and to the next IWC Committee meetings. By 1979, environmental non-governmental organizations (ENGOs) were officially allowed to attend the IWC meetings as observers or even as members of delegations (Nagtzaam, 2009).

In 1982, the issue of the moratorium was revisited – this time with a different result. The IWC's open membership had an important role in how the pro- and anti-whaling controversy has unfolded. Any country can become a member of the IWC by formally adhering to the ICRW.[28] By the beginning of the 1980s, many non-whaling nations joined the organization, including those who had never conducted any whaling-related activities, and each of them had the right to cast a vote. The recruiting of such new members was done by the US and other enthusiastic anti-whaling nations, as well as through aggressive ENGOs' campaigning (Nagtzaam, 2009). In 1982, with 25 votes for, 7 against and 5 abstentions the Commission voted for a ten-year moratorium on commercial whaling by setting the catch limits for all whale species and stocks to zero, which became effective from the 1985–1986 whaling season.[29] The measure was supposed to be a temporary one, explained by uncertainties about the abundance and health of stocks. However, it is still in place now, more than 35 years later.

Although the number of votes "for" the moratorium suggests large support for conservationist or even preservationist agendas, the debates did not end in 1982. Immediately, Japan, along with Norway, Peru,[30] and the USSR, lodged objections to the moratorium (Ellis, 1999).

Japan insisted that the moratorium was not scientifically justified – as certain species and stocks could be targeted sustainably – and Tokyo has kept to this line of argumentation ever since. Representatives of Japan were also dissatisfied with the participation in the IWC of nations, organizations, and individuals that were previously uninterested in whaling and inexperienced in the issue (Ellis, 1999). This discontent turned into Japan following suit – soon after, pro-whaling forces also started employing the tactic of recruiting non-whaling states as supporters of their position. Eventually Tokyo agreed to abide by the moratorium – some suggest that the threat of trade sanctions from the USA has played a detrimental role in this. However, Japan has always advocated for the resumption of commercial whaling.

One tool Japan has been using in its mission to restart whaling is the country's relatively large-scale scientific research whaling programs. Issuing permits for research whaling is allowed by Article VII of the IRWC (ICRW, 1946). The Japanese research whaling programs, including JARPA, JARPA II and NEWREP-A in the Antarctic and JARPN and JARPN II, in the North Pacific, have been provoking controversial reactions nationally and even more so internationally for decades, with mixed opinions on the necessity of the research Japan conducts, its methods, and results (Clapham et al., 2003). Regardless, Japan remained protective of this practice in hopes that it would eventually lead the country and its allies to winning the debate at the IWC. This, however, never happened, and on December 26, 2018, Japan announced its withdrawal from the ICRW and ended its membership in the IWC on June 30, 2019.

Japan leaving this international organization was a bold diplomatic move – although countries permanently withdrawing from international agreements is an option, it is far from being an everyday occurrence and is usually provoked by strong political confrontations. This decision had been called for by the Japanese whaling community for many years (personal observations, 2017–2019), but the country's leadership was hesitant. Japan finally announcing its withdrawal from the ICRW and IWC supports the observation that the issue of whaling is one where Japan is not ready to concede to the anti-whaling forces, even at the cost of damaging its international reputation. Moreover, Japan attended the latest 68th IWC meeting in October 2022 in the status of an observer. This was likely at least partially motivated by Tokyo's intention to show that the resumption of commercial whaling by Japan was not the only objective this country pursued. Japan is holding on to its leadership in the pro-sustainable whaling camp. Close to 40 other nations members of the IWC support the idea of whaling, and Japan was at the forefront of this group. Additionally, Japan still wants to campaign for the principle of sustainable use, claiming that the issue is larger than whaling alone and that anti-whaling is just one case of Western imperial environmentalism – when developed nations prescribe what resources are to be consumed and in what way, the rest of the world is to follow regardless of their circumstances (personal communication, March 2023). Both of these points will be revisited in Chapter 5.

Immediately following Japan's withdrawal from the ICRW and IWC, the country resumed commercial whaling in its territorial waters and exclusive economic zone on July 1, 2019. The outcomes of this move are still to be seen, as it has only been several years. Moreover, for most of this time, Japan, along with the rest of the world, was dealing with the COVID-19 pandemic and its consequences, which impacted the whaling industry as much as it did every other human activity.

What can be said for now is that although the whaling industry in Japan is not the most lucrative one at the moment, it is certainly still alive. The community who upholds it is also still motivated, despite the challenges they recognize, including the aging of the workforce, difficulties attracting young professionals due to hard working conditions at sea and uncertain prospects on land, and the loss of mass consumers' interest because of how long whale meat was expensive and relatively rare (personal communication, 2021). More on all that is to come in Chapter 5 as well.

Notes

1 See glossary for definition.
2 Interconnectedness in the context of the whaling issue will be further discussed in Chapter 5.
3 See, for example, Flipper (1964–1967) – a well-known television show featuring a dolphin as one of its main characters; see, for example, Rothenberg (2008), for an account of the inception of scientific and popular interest in whale songs.
4 These included blue whale – the largest mammal on earth, right whale, Humpback whale, Fin whale, Sei whale, and Sperm whale, certain stocks of grey whale and the bowhead whale.
5 According to a 1965 assessment for the Red List of Threatened Species by the International Union for Conservation of Nature, these species of whales were considered "very rare and decreasing in numbers", or "less rare but believed to be threatened". Further assessments gave them the status of endangered, which six of the eight species still hold now (IUCN, 2015).
6 Some of the IWC member countries are whaling nations; some are former whaling nations, but most have never had whaling industries and joined through the diplomatic efforts of either the pro-sustainable whaling or the anti-whaling camp to support the respective side of the debate.
7 This can be found at www.whaling.jp/english/history.html
8 The definition and history of this category of whaling are detailed below.
9 For a more detailed discussion of the components of the Japanese whaling scene – its actors, their functions, and the relations between them – see Chapter 5 on the social environment and landscape of Japanese whaling.
10 Can be accessed on MOFA's website, the "Management of Cetacean Resources" section: https://www.mofa.go.jp/policy/economy/fishery/whales/iwc/chronology.html
11 It can be accessed at www.whaling.jp/english/history.html
12 Printed copies are available at the offices of the Institute of Cetacean Research in Tokyo, Japan, as well as during the many whaling-related events around the country. It can also be found on the organization's website in the section with brochures' archives: https://www.icrwhale.org/pdf/04-B-ken.pdf

13 I was first acquainted with the beauty of Taiji whaling boats on my tour around Taiji Whale Museum, where I was kindly guided by its deputy director, Tetsuo Kirihata, in July 2017. The museum's main exhibition hall features a life-size model of a *sekobune*, as well as pictures of other boats with detailed explanations. There is now a digital version of the museum that features a page on the styles and designs of Taiji whale boats. See it here: https://kujira-digital-museum.com/en/categories/14/articles

14 A radical is a core building block of a character, often giving a sense of the character's overall meaning.

15 A quadrillion is 10^{15} or 1,000,000,000,000,000.

16 The same linguistic point applies to dolphins. Although currently it is most common to write the word "dolphin" in *katakana* alphabet – イルカ – the character for this word also includes the same "fish" radical – 鯆.

17 This point was mentioned in Ellis, 1999; Komatsu & Takagi, 2006; personal communication with one research participant.

18 The current status of whaling in Taiji and the experiences of the town's representatives with whaling and anti-whaling will be discussed in detail in further chapters.

19 Small-type coastal whaling (SCTW) and large-type coastal whaling (LTCW) are categories of whaling developed and regulated by the Japanese government and conducted uniquely in Japan. See Glossary for more details.

20 That whaling operation occurred in a different location – off the coast of Hokkaido in the north of Japan, and whaling boats from different companies were invited to take part.

21 Sanriku is a historical name of a large area on the northeastern side of the Japanese Honshu island that corresponds to the modern territory of Aomori and Iwate Prefectures, as well as parts of Miyagi Prefecture including Ayukawa. Sanriku-kaigan (Sanriku Coast) is also a colloquially used title, as well as Sanriku-oki that denotes a part of the Pacific Ocean around Sanriku.

22 This word combination is often used to describe Ayukawa, as well as other Japanese towns with a long whaling history. The website of Ishinomaki city, which Ayukawa is now a part of, explains that Ayukawa is known as a "*kujira no machi*" all around Japan. The page also has a downloadable brochure on the history and modernity of the port, which is called "*Kujira no machi Ayukawa: Kindai hogei 100 shunen*" ["Town of Whales Ayukawa: Around 100 years of Modern Whaling"] (2006).

23 Both in Wadaura and Ayukawa the main whale species targeted during those years was Baird's beaked whale. Another species that was hunted in Ayukawa is northern form short-finned pilot whale. As mentioned throughout this book, small species of cetaceans do not fall under the IWC regulation and the catch quotas for those were issued on the national level. To give an idea about the actual numbers of small species of whales taken in Japan, in 2018 the annual quota for Baird's beaked whale was 66 whales for the whole country, with 52 allocated for the Pacific coastal line whaling towns and shared equally between Wada and Ayukawa, 10 for the Sea of Japan and caught by whalers from Hakodate, and 4 for the Sea of Okhotsk and the catch was administered from a whaling station in Abashiri. In the same year the annual quota for southern form short-finned pilot whale was 50, which was shared between Wadaura and Taiji. The annual quota for northern form short-finned pilot whales was 50 and was taken from Ayukawa. All of the quotas were reviewed periodically but did not drastically deviate from these numbers.

24 Curiously, the right whale was given its name by whalers, and it was because it was considered the "right" whale to catch. It moves at a relatively low speed of up to 12 km/h and its body floats after death because of a thick layer of blubber, as opposed to other species that sink. It also has very long strong and flexible baleen – up to 4 meters in the Greenland right whale, which was widely used in production of many commercial products, including instruments' strings, corsets and toys.

25 The 1982 United Nations Convention on the Law of the Sea (commonly referred to as UNCLOS) established the norm of 12-nautical-mile territorial waters, and most countries follow it now.

26 A number of scholars agree with this assessment in their works, for example, Andresen, 1993; Christol, Schmidhauser and Totten, 1976; Nagtzaam, 2009; Oberthür, 1999; van Drimmelen, 1991.
27 Whaling along with hunting or harming other species of marine mammals was made illegal in the USA in 1972, when the Congress passed the Marine Mammal Protection Act. Notably, that is well before the moratorium on commercial whaling went into effect in 1986. However, the whaling industry was in sharp decline in this country in the first decades of the 20th century.
28 See the IWC website "Membership and Contracting Governments" https://iwc.int/members
29 See the IWC website "Commercial Whaling" https://iwc.int/commercial
30 Peru withdrew the objection soon after.

References

Abdel-Hadi, A., Tolba, M. K., & Soliman, S. (2010). *Environment, health and sustainable development*. Toronto: Hogrefe.

Andresen, S. (1993). The effectiveness of the International Whaling Commission. *Arctic, 46*(2), 108–115.

Christol, C. Q., Schmidhauser, J. R., & Totten, G. O. (1976). The law and the whale: Current developments in the international whaling controversy. *Case Western Reserve Journal of International Law, 8*(1), 149–167.

Clapham, P. J., Berggren, P., Childerhouse, S., Friday, N. A., Kasuya, T., Kell, L., & Brownell, R. L. Jr. (2003). Whaling as science. *BioScience, 53*(3), 210–212.

Clark, C. W., & Lamberson, R. (1982). An economic history and analysis of pelagic whaling. *Marine Policy, 6*(2), 103–120.

Day, D. (1992). *The whale war*. New York, NY: Harper Collins Publishers.

Ellis, R. (1999). *Men and whales*. Guilford: Globe Pequot Press.

Endo, A. (2011). Henyo suru geirui shigen no riyo jittai: Wakayamaken Taijicho no shokibo engan hogei wo jirei to shite [The state of utilization of the changing cetacean resources: The example of Taiji, Wakayama Prefecture]. *National Museum of Ethnography, 97*, 237–269.

Fackler, M. (2011, March 24. Japanese town mulls future without whaling industry. *The New York Times*. Accessed online https://www.nytimes.com/2011/03/25/world/asia/25whale.html

Gambell, R. (1993). International management of whales and whaling: An historical review of the regulation of commercial and aboriginal subsistence whaling. *Arctic, 46*(2), 97–107.

Heazle, M. (1993). Japan: A community on edge – Ban on whaling threatens future of Ayukawa. *Far Eastern Economic Review, 156*(23), 34–35.

ICRW. (1946). International Convention for the Regulation of Whaling. Accessed online https://archive.iwc.int/pages/view.php?ref=3607&k=

Ishinomaki City. (2006). Kujira no machi Ayukawa: Kindai hogei 100 shunen [Town of Whales Ayukawa: Around 100 years of Modern Whaling]. Accessed online www.city.ishinomaki.miyagi.jp/cont/10151000/1000/2338/200608_02_03.pdf

IUCN. (2015). Balaenoptera musculusssp.Intermedia. The IUCN Red List of Threatened Species. Accessed online http://www.iucnredlist.org/details/41713/0

IUCN SSC. (2016). Status of the world's cetaceans. Cetacean Specialist Group. Accessed online http://www.iucn-csg.org/index.php/status-of-the-worlds-cetaceans/

IWC. (n.d.). Commercial whaling. Accessed online https://iwc.int/commercial

IWC. (2022). The IWC Organisational Chart. Accessed online https://archive.iwc.int/pages/view.php?ref=6372&k=

Japan Small-Type Whaling Association. (2002). Japanese community-based minke whaling. Accessed online www.jstwa9.com/hp-eng/pdf/2002E.pdf

Japan Whaling Association. (1987). *Whale and traditions of diet*. Tokyo: Japan Whaling Association.

Kalland, A., & Moeran, B. (1992). *Japanese whaling: End of an era?* Copenhagen: Scandinavian Institute of Asian Studies.

Kita, Y. (2007). The four hundred years of Taiji whaling and the years to come. Report and Proceedings: The 5th Summit of Japanese traditional Whaling Communities. Taiji: Taiji Town Office.

Komatsu, M., & Takagi, Y. (2006). *Des baleines et des hommes [Whales and men]*. Tokyo: The Institute of Cetacean Research.

Morishita, J., & Goodman, D. (2005). Role and problems of the Scientific Committee of the International Whaling Commission in terms of conservation and sustainable utilization of whale stocks. *Global Environmental Research*, *9*(2), 157–166.

Nagtzaam, G. J. (2009). The International Whaling Commission and the elusive great whale of preservationism. *William & Mary Environmental Law and Policy Review*, *33*(2), 375–447.

Nicol, C. W. (1979a) Ayukawa, a whaler's town. *Whales and Whalers*. Tokyo: Japan Whaling Association.

Nicol, C. W. (1979b). Taiji - winds of change. Tokyo: Japan Whaling Association.

Oberthür, S. (1999). The International Convention for the Regulation of Whaling: From overexploitation to total prohibition. *Yearbook of International Co-operation on Environment and Development 1998/99*. Accessed online https://www.researchgate.net/publication/242125997_The_International_Convention_for_the_Regulation_of_Whaling_From_OverExploitation_to_Total_Prohibition

Rothenberg, D. (2008). Whale music: Anatomy of an interspecies duet. *Leonardo Music Journal*, *18*, 47–53.

Ruffle, A. M. (2002). Resurrecting the International Whaling Commission: Suggestions to strengthen the conservation effort. *Brooklyn Journal of International Law*, *27*(2), 639–671.

Scientific Committee Handbook. (2018). Working methods of the IWC's Scientific Committee. Accessed online https://archive.iwc.int/pages/view.php?ref=7670&k=

Skodvin, T., & Andresen, S. (2003). Nonstate influence in the International Whaling Commission, 1970-1990. *Global Environmental Politics*, *3*(4), 61–86.

Taiji Declaration on Traditional Whaling. (2007). *Report and proceedings: The 5th Summit of Japanese traditional whaling communities*. Taiji: Taiji Town Office.

Takahashi, J., Kalland, A., Moeran, B., & Bestor, T. C. (1989). Japanese Whaling culture: Continuities and diversities. *Maritime Anthropological Studies*, *2*(2), 105–133.

Tani, K. (2012). Shouchiikibetsu ni mita hiashi nihon daishinsai hisaichi ni okeru shibousha oyobi shibouritsu no bunpu [Distribution of the number of deaths and mortality rate in the areas affected by the Great East Japan Earthquake by region]. Saitama University, Faculty of Education, Department of Geography, 32, 1–26.

Tønnessen, J. N., & Johnsen, A. O. (1982). Translated by R. I. Christophersen. *The history of modern whaling*. Berkeley, CA and Los Angeles, CA: University of California Press.

van Drimmelen, B. (1991). The international mismanagement of whaling. *Pacific Basin Law Journal*, *10*(1), 240–259.

2 Historical and contemporary take on whaling

To place this work in the context of the already existing resources on the phenomenon of whaling, a combination of chronological and conceptual approaches is used in this chapter. First, there will be an overview of what is written on the issue from a historical perspective, followed by an introduction of the more recently emerged trends in the discussions of this issue. Many scientific disciplines have concerned themselves with whaling, including aquatic sciences and oceanology (marine biology), ecology, conservation biology, environmental science, history, anthropology, philosophy (ethics), economics, law, politics, international relations, etc. Of more obvious relevance to this research project are publications centered around international relations and politics, as well as anthropology and interdisciplinary work that includes these angles. I will mainly focus on these here.

Documenting whaling from a historical perspective

Although whaling has become of wide interest to the general public only in the past four decades or so, this practice has been around for much longer, first as a subsistence activity and later as a commercial one. According to some recent archaeological discoveries, humans were reportedly already whaling around 6000 BCE (BBC, 2004), or as more established evidence shows at least as early as 2000 BCE (Marrero & Thornton, 2011), and although the exact time of its start is still debated and varies across regions and countries (Estes, DeMaster, Doak, Williams, & Brownell, 2006), it can be said with certainty that the history of this practice is long and complex.

There is relatively little information about whaling in prehistoric to medieval times, however, there are some works covering what is known about it. For example, the first five parts of the lengthy introduction by Savelle and Kishigami to the 2013 volume "Anthropological Studies of Whaling" cover this period and so do books on the history of whaling in general, such as "Whaling World: From Prehistoric Times to the Present Day" by Euller (1970), "Whaling in the North Atlantic: From Earliest Times to the Mid-19th Century" by Proulx (1986).

There are also works featuring a specific aspect of ancient whaling. For instance, an article by Douglas, Smol, Savelle, and Blais (2004) on prehistoric Inuit whaling uncovered a previously unrecognized negative effect on the freshwater

DOI: 10.4324/9781003255031-3

ecosystems in the Arctic. Another article of this kind by Krupnik and Kan (1993) analyzes prehistoric Eskimo whaling in the Arctic. Let us review this latter one in more detail, as the authors attempt to analyze seemingly conflicting views on the practice of whaling during one particular time and in one place gives us a good idea of how much perspectives differ when it comes to this subject overall.

Krupnik and Kan studied the practice of targeting calves and young whales rather than adult animals by prehistoric indigenous people. They attempt to find the middle ground between the view that such "slaughter of calves" was a disruptive activity from the ecological perspective and the opinion that this was an early purposeful attempt at managing the whales' population. Defining his argument as a compromise between the two opposing views, the authors conclude that for Eskimo whalers hunting calves was a time and effort-saving pragmatic strategy. While Krupnik and Kan's inquiry is set in prehistoric times, this analysis of a distant-in-time subsistence whaling practice in a way echoes the deliberations on the nature of whaling today. What is the purpose of it and what should the attitudes to it be? Krupnik and Kan reviewed archeological data, in our case we have the data from current catches, but what to make of it all remains up to interpretation. Is it "barbarous" or just "pragmatic" or even sustainably minded? While Krupnik and Kan find no evidence of "ecological mentality" per se in his case, the very fact that it is "highly debated" and the author is trying to find a middle ground, or a more nuanced stance, is of interest to this book. The juxtaposition of a "predatory" attitude and an "eco-friendly" attitude toward the environment is a recurrent theme in discussions on whaling, its management, protection of species, and reasons some countries are still whaling while others are not. This duality, however, is socially constructed and is too general to recognize the specificity of different peoples, spaces, and circumstances. In whaling, as in many other socio-environmental issues, the reality is more complex than that.

Let's review some whaling-related work that dives into the anthropological side of this practice. In a chapter by Sheehan (1985) – "Whaling as an organizing focus in Northwestern Alaskan Eskimo Society" – and a book by Creighton (1995) – "Rites and passages: The experience of American whaling, 1830–1870." – whaling is seen as an activity that shapes people's lives and contributes to human development. The first work explores the significant positive role whaling played in the development of social complexity in prehistoric Northwestern Alaskan Eskimo society. The second, a lengthier one, focuses on a different period – from 1830 to 1870, but also looks at whaling from the perspective of people who were engaged in it. It depicts the impact of whaling on seafarers, or whalemen, during the period that the author identifies as the peak of the industry's development in the United States of America. Such a take on whaling, where it is seen as people's source of income, main occupation, and an important part of their life, is valuable in the light of the modern international community's negative attitude to both the practice and countries and people who engage in it. Although the so-called aboriginal subsistence whaling is recognized as essential in supporting some communities' nutritional needs and is allowed by the Schedule to ICRW, in some cases this concept is also criticized and debated, as discussed by, for example, Caulfield (1993),

Hamaguchi (2013) and Khoury (2015). The argument about whaling being a formational and integral part of some communities' culture, including some regions in Japan, is also often brought up in modern days whales' related debate, which will be discussed further.

Some interesting non-academic works on whaling belong to the category of non-fiction and give accurate historical accounts of whaling-related events, but also romanticize the issue, turning whaling into an epic drama, where men are facing the dangers of the sea and its largest inhabitants. An example of such work is "Leviathan: The History of Whaling in America" by Dolin (2007). Although works like this often go unnoticed by academic research, in the particular case of whaling-related arguments they represent an important idea that the subject in question is highly prone to emotional interpretations. Initially, the narrative was built around the people who engage in whaling, and their work is seen as brave and full of adventures. Later on, with the emergence of the anti-whaling discourse, it is the targeted animals who were put at the center. They became the friendly and intelligent gentle giants that whales are often seen as now. Emotive narration style in relation to whaling has been developed first in the popular media in the late 20th–early 21st century and has since penetrated scientific and policy discourse as well.

An important period in whaling is associated with the start of the Basque people's activity. It is amply documented in "Men and Whales" by Ellis (1991), which commences with descriptions of whaling by the Basques starting from around the 10th–11th centuries CE. While pre-Basque whaling was a subsistence activity, the Basque people of the Pyrenees region, between the modern territories of France and Spain, started whaling commercially, trading with other countries, which paved the way for whaling into the international arena. The Basque whalers pioneered many new techniques, were considered the best in this occupation, and dominated it for nearly five centuries. Such a significant role of these people in the development of whaling resulted in works dedicated solely to them and their craft – Edvardsson and Rafnsson (2006), for instance. During the Basques period, there was already some competition over whaling grounds, and with time the Basques reached as far as Newfoundland, the Faroe Islands, and Iceland in search of catch. The Basque whaling slowly weakened mostly due to the decimated stocks of the species they were mainly targeting – the right whale and the bowhead whale.

Documenting modern whaling

As discussed in the previous chapter, the term "modern" marks an important shift in the history of whaling, and it is associated not only and not as much with a specific time in history, but with then-novel methods employed for whaling, many of which are still in use now. Confirming the critical importance of this stage of whaling history, a large number of studies start their analysis from this period – the late 19th–early 20th centuries.

The history of modern whaling is well documented in several works. "The History of Modern Whaling" is one of the most comprehensive works on the subject. It is a volume by Tønnessen and Johnsen (1982), which is the translated and abridged version of the four-volume Norwegian original published between 1959 and 1970. It began around 1860 when the newly developed techniques such as different types of explosive harpoons and faster catcher boats enabled whalers to target bigger species of whales, including the largest mammal on Earth – the blue whale. Tønnessen and Johnsen's work, which takes us to the end of the 1960s in the history of whaling, is neutral in style having no negative or positive overtones. This highlights the fact that it is only with the development of strong public interest toward cetaceans and the growing environmentalism in the late 1960s and early 1970s[1] – which is after Tønnessen and Johnsen's book was published – that the attitudes to whales and the pertinent discourse have changed from seeing them as a manageable recourse, that was not managed well enough for several reasons, toward depicting them in a dramatic way and developing protectionist feelings for them that do not reflect the state of whale stocks.

Many researchers of the topic agree with the fact that the next important shift in the history of whaling is the development of pelagic whaling, which was what eventually brought certain species of whales close to extinction. Pelagic whaling is the focus of quite a few research projects, including the works of Surrency (1964), Sommers (1967), Clark and Lamberson (1982), and Bockstoce (1995).

The articles by Surrency (1964) and Sommers (1967) date back to the 1960s – when the industry was facing an increasing depletion of stocks, but right before the rise of environmentalism in the 1970s and before the moratorium on whaling was introduced in 1982. Both authors express concern about the poor state of whale stocks – Sommers talking about the Antarctic in particular, while Sommers refers to the problem on the global scale – and both state that the international efforts to tackle the issue had not been effective enough, although by the time of their writing, the IWC was established and catch quotas for certain species and stocks of whales were already introduced. As the authors identify and analyze the problem, they focus on the difficulties the countries had in managing a global natural resource and the need of devising better ways of international control and cooperation in this resource's utilization. Despite the reduced numbers of whales, their survival and increasing stocks to "a more economical level", as Surrency puts it, is seen as instrumental in serving the needs of people, including achieving food security for the growing human population.[2] Whales are not seen as special animals, deserving treatment any different from other species, no mention is made of them having exceptional intellectual abilities, feelings, particularly appealing appearance or unusual relations with people. Depicting whales in that way was not common before the rise of environmentalism in the 1970s, however, presently the popular discourse routinely describes these animals in anthropomorphic terms (Kalland, 1993a).

In an article on the economic history of the whaling industry Clark and Lamberson (1982) also focus on pelagic whaling. The authors test the theory of common resources and compare it to the optimal resource exploitation concept, ending up criticizing both for not capturing the actual whaling situation – the first not accounting for the fact that the competition in whaling was not unrestricted (only a relatively small number of nations whaled) and not entirely unregulated (the IWC did exert some impact) as the theory predicts; and the second – for being unrealistic as the maximum sustainable yield it seeks is extremely prone to fluctuations and thus difficult to define. This work's importance is not only in its original conceptual framing but also in its concise and accurate account of the regulation of whaling and its elements.

Clark and Lamberson (1982) are among the many authors who agree that of all elements forming the entirety of the global whaling regime, perhaps, the most important is the international organization that has been the pillar of its functioning throughout the past seventy decades – the IWC. In some written accounts, in fact, the evolution of the whole whaling regime is equated with the developments this organization underwent in the years of its existence. One of the most prominent examples of this is a three-volume discussion of the yearly proceedings of the IWC by Birnie (1985). It is one of the fullest available accounts of the work of the IWC, the only inconvenience related to it being the fact that since the work was published in 1985 quite a few important developments have emerged in whaling regulation since then. A more recent take on the role of IWC is discussed by Peterson (1992), who pays more attention to how the emergence and strengthening of environmental discourse and activism influenced this organization and the changes to its agenda.

In light of the early inability of the IWC to prevent the collapse of certain stocks and the whaling industry and later on the debates on the necessity of the ban on commercial whaling introduced in 1982, a great deal of research is centered on the IWC's effectiveness. Coming up to the moratorium and up until the latest 68th meeting held in 2022, the organization's annual and then biennial meetings epitomized a stalemate in environment-related discussions – the pro-sustainable whaling nations and those who support the moratorium on commercial whaling have held their ground and neither team's position have changed significantly in the past decades. As Juma (2008) puts it, polarization came to be the norm in the IWC process. The organization was facing major challenges and its operations were receiving harsh criticism (Andresen, 1993; Burns & Wandesforde-Smith, 2002). The need for the IWC to be more flexible and prompt in responding to the changing international norms is also noted by some scholars (Juma, 2008).

The IWC's difficulties in performing the tasks of both developing the industry and preventing harm to whale species and the often incompatible nature of these two objectives is discussed by Harris (2005). The author brings up the concept of a "false positive", which is a result of an unnecessary environmental precautionary step. While it might seem that such steps are yielding positive results initially, especially if examined from the perspective of environmental protection, in some cases, they are later found unnecessarily strict and thus preventing profit generation and industry development. The pro-sustainable whaling nations trying to prove that the existing rigid rules around the preservation of some whale species is exactly one such unnecessary environmental precautionary step. They employ this argument when advocating for the resumption of commercial whaling, stating that some species of whales can be taken in reasonable numbers without harm to the whole population's survival.

Most recently, with Japan having left the IWC, the 2022 68th meeting of the organization was, according to Morishita (2023) "less confrontational", as more effort was spent on discussions of programs that would protect whales than on pro- and anti- whaling debates. It remains to be seen, however, what the future of this organization will look like, as it struggles with budgetary issues in the face of insufficient financial support from members and continuing opposition from the member countries who support pro-sustainable whaling, especially developing nations (Ibid.)

Research on Japanese whaling

Japan is not the only nation that continues to conduct whaling. And there are people who simply criticize whaling – in Japan or elsewhere, regardless of its purposes, condemning the act itself and considering the nations and people who still whale cruel and misguided (Darby, 2008). However, Japan's whaling arguably attracts the most international criticism (Ackerman, 2002; Komatsu & Takagi, 2006), regardless of the fact that it did not whale commercially from the time the moratorium went into force and until 2019, thus, conforming to the IWC regulations.

The faultfinders mostly confronted Japan's whaling under special permits, or as they are widely known, research whaling programs, that the country had been conducting in the high seas of Antarctica and the North Pacific while it was a member of the IWC. Japanese research on whaling was often proclaimed unscientific in scholarly articles and commentaries (Ackerman, 2002; Papastavrou, 2006; Clapham, 2015; Brierley and Clapham, 2016). Some authors, mostly in media, including The Washington Post, BBC, CNN, etc., and publications from wildlife protection organizations, such as WWF and IFAW, put the word research in quotation marks when talking about the issue, to show their attitude and express the belief that this was only a façade covering commercial activity since the by-products of the Japanese research programs – namely, whale meat – was sold in the country's markets (Feltman, 2015; BBC, 2015; Whiteman, 2015).

However, statements equating research to commercial activity in this case lack basis. Commercial activities are those undertaken primarily to make a profit. Japanese research whaling by-products were, indeed, sold. But they were never marketed as regular products, their price was not determined by the market, and the revenues from these sales were immediately reinvested in research. Hence, making a profit could not have been considered the purpose of these transactions and it would be inaccurate to call Japanese research commercial in nature. Additionally, selling by-products of research is allowed and even encouraged by the ICRW. Having nutritiously valuable products consumed, instead of simply discarding them is also sustainable.

Constructive criticism of research design, data collection process, analysis, or conclusions is a standard process in scientific research. There was plenty of that at the IWC SC and other specialized bodies and journals when it came to the results of Japan's research programs. Joji Morishita, Japan's commissioner to the IWC and the organization's chairman in 2016–2018, assures that Japan welcomes constructive scientific dialogue on the matter (personal communication, 2016). However, when it comes to popular publications, their authors would often jump to conclusions without explaining exactly which aspects of the Japanese research programs lacked scientific rigor and accuracy, making the critics' argument appear weak and lacking merit.

It is also important to reiterate that Japan never concealed the fact that it intended to resume commercial whaling, and was doing everything to attain that goal, including research programs. Former Prime Minister Shinzo Abe stated on several occasions that Japan would continue research whaling to "revive commercial whaling" (Lies, 2014), and the official documents on the Japanese position on whaling for a long time admitted the possibility of controlled commercial whaling resumption (The Japanese Government's Position on Whaling). So again, the Japanese government had never been equivocal about the main objective of its post-moratorium and pre-IWC withdrawal whaling policy. With Japan's withdrawal from the organization and the resumption of commercial whaling of certain healthy whale stocks from July 1, 2019, the country also ceased its research operations outside of its territorial waters and Exclusive Economic Zone. With this, the issue of Japanese research whaling will, perhaps, be attracting less attention. But the response of the international community to these newest developments remains to be seen.

Since stocks of certain species[3] of whales are known to be abundant enough to be taken sustainably, Japan's stance that the industry, if managed well, could be both useful and profitable seems straightforward and not need additional explanation. However, the global media and scientific community have shown considerable preoccupation with the question of the reasons behind Japan's continuous effort to keep whaling. There are several predominant opinions on Japan's motivations – first and foremost, there are those who hold that the roots of the government support of this practice lie in the fact that whaling is a part of Japanese culture, or, as some specify, culinary culture. As it is an important cultural phenomenon, giving it up would mean letting go of a part of Japanese heritage. This book elaborates on what areas of Japan have been involved in whaling the most and how Japanese whaling and whale meat-based cuisine are different from any others. Among the proponents of the cultural significance argument are Takahashi, Kalland, Moeran, and Bestor (1989), who analyze different types of whaling in pre-historic to modern times in Japan and argue that an "integrated whaling culture" does exist in this country. One of the article's authors – Arne Kalland was a prominent pro-whaling scholar who not only argued for the cultural right of Japan and his native Norway to whale but also suggested that whale totemization and anti-whaling is not beneficial and even harmful to the environment, as it serves an easy cause for governments to join to pretend to be "green" (1993a, 1993b). Other authors mentioning the importance of Japanese culture and even religion as the context in which Japan has rejected the anti-whaling norm include Freeman (1998), Danaher (2002), Komatsu (2005), etc. The Japanese government is also supportive of the cultural justification of their policies (The Japanese Government's Position on Whaling), however, Joji Morishita and a few other individuals who were involved in the whaling debates for a while are somewhat cautious about over-stating this line of argumentation.

They worry that the "culture" argument might make developing whaling industries in places with no such culture difficult. And that contradicts Japan's belief that whaling could potentially benefit many developing countries in terms of food security, including those who have no tradition of whaling for now (personal communication, November, 2016).

Blok (2008) brings up the concept of identity and suggests that the Japanese pro-sustainable whaling stance is a manifestation of the need to protect this identity. At the same time the culture and identity argument is deemed a political construct by other authors (Ishii and Okubo, 2007) and the existence of Japanese whaling culture as such is also questioned (Watanabe, 2009). The whaling culture denialists argue that the Japanese government is mostly reacting to the interplay of the country's political groups and elites whose interests are related to the whaling industry, or securing marine fisheries in general – Japan failing to successfully negotiate on the issue of whaling might weaken its position when it comes to marine resources overall, therefore the country's government continues to insist (Kedzlie, 2014; Morikawa, 2009; Sekiguchi, 2007). The argument of the economic profit or its possibility in the future is also explored by some authors, although most agree that for years Japanese whaling was generating only loss, and predict that this situation would not change foreseeable future (Hirata, 2005; Kedzlie, 2014). Again, Japan's withdrawal from the ICRW and IWC in 2019 could be followed by significant changes. However given the influence of the COVID-19 pandemic and the fact that 2022 was only the fourth commercial whaling season for Japan, it is too early to tell with certainty where the industry is headed. We will look at some preliminary impressions and forecasts further in Chapter 5.

Going back to the reasons behind Japan's insistence on whaling there are also those, who tend to think that no one factor can explain this, and all of the mentioned above aspects in some way influence the outcome we currently witness (Catalinac & Chan, 2005; Hirata, 2005; Morishita, 2006). To add to the range of discussions on the rationale of Japan's whaling, there is also an opinion that not only the act of whaling per se should be criticized, but also the inability of the Japanese government to properly explain their actions to the international community (Cortazzi, 2015).

Taking into due consideration the material on whaling in general and whaling in Japan that is already out there, in this book, I attempt to find an alternative to the general direction of the academic and popular discourse on the matter. Many curious minds focused on the reasons behind Japan's past research programs and the fact that Japan still whales. This book suggests that before asking "why" we should understand the "who" better – focus on the human factor behind the issue of whaling. Personal perspectives of those whose day-to-day activities are related to whaling can give us a better understanding of the issue and, perhaps, the "why" will then become not as puzzling.

Notes

1 See, for example, Abdel-Hadi et al. (2010) on the development of environmentalism in the 1970s; and Babcock (2013) for mentions of the role of whales in the advance of global environmentalism. Both points are also mentioned further in this book.

2 The idea of whaling contributing to a country's food security is a prominent one in the Japanese pro-sustainable whaling discourse now, but it is viewed with skepticism in the West. See more on that in Chapter 5.

3 As of 2023, Japan targets minke, Bryde's and sei whale in its territorial waters and EEZ. In 2023 the total allowable catch (TAC) for these was set at 136 minke, 187 Bryde's, and 25 sei whales. To calculate TAC, first the quotas are calculated using the method adopted by the IWC, then the average number of whales of these species caught as bycatch in fishing nets during the previous five-year period is subtracted (the data from the Ministry of Agriculture, Forestry and Fisheries is available online https://www.jfa. maff.go.jp/j/whale/attach/pdf/index-54.pdf). There are competing theories about the abundance of sei whale population in this area, but both minke and Bryde's whale are not considered endangered.

References

Abdel-Hadi, A., Tolba, M. K., & Soliman, S. (2010). *Environment, health and sustainable development*. Toronto: Hogrefe.

Ackerman, R. (2002) Japanese Whaling in the Pacific Ocean: Defiance of International Whaling Norms in the Name of "Scientific Research," Culture, and Tradition, 25 B.C. International and Comparative Law Review, 323–334.

Andresen, S. (1993). The effectiveness of the International Whaling Commission. *Arctic, 46*(2), 108–115.

BBC (2004, April 20). Rock art hints at whaling origins. Accessed online http://news.bbc. co.uk/2/hi/science/nature/3638853.stm

BBC (2015, November 28). Japan to resume whaling in Antarctic despite court ruling. Retrieved from http://www.bbc.com/news/world-asia-34952538

Birnie, P. (1985). International regulation of whaling: From conservation of whaling to conservation of whales and regulation of whale-watching. Vols. I and II. New York, NY, London, Rome: Oceana Publications, Inc.

Blok, A. (2008). Contesting global norms: Politics of identity in Japanese pro-whaling countermobilization. *Global Environmental Politics, 8*(2), 39–66.

Bockstoce, J. (1995). *Whales, ice, and men: The history of whaling in the Western arctic*. Seattle, WA: University of Washington Press.

Brierley, A., & Clapham, P. (2016). Japan's whaling is unscientific. *Nature, 529*, 283.

Burns, W., & Wandesforde-Smith, G. (2002). The International Whaling Commission and the future of cetaceans in a changing world. *Review of European Community & International Environmental Law, 11*(2), 199–212.

Catalinac, A., & Chan, G. (2005). Japan, the West, and the whaling issue: Understanding the Japanese side. *Japan Forum, 17*(1), 133–163.

Caulfield, R. A. (1993). Aboriginal subsistence whaling in Greenland: The case of qeqertarsuaq municipality in West Greenland. *Arctic, 46*(2), 144–155.

Clark, C. W., & Lamberson, R. (1982). An economic history and analysis of pelagic whaling. *Marine Policy, 6*(2), 103–120.

Cortazzi, H. (2015, December 7). Japan has little to gain by resuming its whale hunt. The Japan Times. Accessed online https://www.japantimes.co.jp/opinion/2015/12/07/commentary/japan-commentary/japan-has-little-to-gain-by-resuming-its-whale-hunt/

Creighton, M. S. (1995). *Rites and passages: The experience of American whaling, 1830–1870*. New York, NY: Cambridge University Press.

Danaher, M. (2002). Why Japan will not give up whaling. *Pacifica Review: Peace, Security & Global Change, 14*(2), 105–120.

Darby, A. (2008). *Harpoon: Into the heart of whaling*. Cambridge: Da Capo Press.

Dolin, E. J. (2007). *Leviathan: The history of whaling in America*. New York, NY: W. W. Norton & Company.

Douglas, M. S., Smol, J. P., Savelle, J. M., & Blais, J. M. (2004). Prehistoric inuit whalers affected arctic freshwater ecosystems. *Proceedings of the National Academy of Sciences*, *101*(6), 1613–1617.

Edvardsson, R., & Rafnsson, M. (2006). Basque whaling around Iceland: Archaeological investigation in Strákatangi, Steingrímsfjörður. Accessed online https://w390w.gipuzkoa. net/WAS/CORP/DBKVisorBibliotecaWEB/pdf/672532?tipoPeticion=ikusi&titulo=Basq ue+whaling+around+Iceland+archeological+investigation+in+Str%C3%A1katangi%2C +Steingr%C3%ADmsfj%C3%B6rour+%2F+%5BRecurso+electr%C3%B3nico%5D&o rria=0&errorImg=true

Ellis, R. (1991). *Men and whales*. Guilford: Globe Pequot Press.

Estes, J. A., DeMaster, D. P., Doak, D. F., Williams, T. M., & Brownell, R. L. (Eds.). (2006). *Whales, whaling, and ocean ecosystems*. Berkeley, CA: University of California Press.

Euller, J. (1970). *Whaling world: From prehistoric times to the present day*. New York: Doubleday & Co.

Freeman, M. (1998). Japanese Community-based whaling, international protest, and the new environmentalism. In D. Myers, & K. Ishido (Eds.), *Japan At the crossroads: Hot issues for the 21st century* (pp. 13–31). Tokyo: Seibundo.

Feltman, R. (2015). A Japanese vessel is set to kill 333 whales for 'research' — but is science really behind the hunt? The Washington Post. Accessed online https://www.washing-tonpost.com/news/speaking-of-science/wp/2015/11/30/a-japanese-vessel-is-set-to-kill-333-whales-for-research-but-is-science-really-behind-the-hunt/

Hamaguchi, H. (2013). Aboriginal subsistence whaling revisited. In N. Kishigami, H. Hama-guchi, & J. M. Savelle (Eds.), *Anthropological studies of whaling* (pp. 59–67). Osaka: National Museum of Ethnology.

Harris, A. W. (2005). *The Best scientific evidence available: The whaling moratorium and divergent interpretations of science. William & Mary Environmental Law and Policy Review*, *29*(2), 375–450.

Hirata, K. (2005). Why Japan supports whaling. *Journal of International Wildlife Law & Policy*, 8(2–3), 129–149.

Ishii, A., & Okubo, A. (2007). An alternative explanation of Japan's whaling diplomacy in the post-moratorium era. *Journal of International Wildlife Law & Policy*, 10, 55–87.

Juma, C. (2008). The Future Of The International Whaling Commission Strengthening Ocean Diplomacy. Special Advisor International Whaling Commission Report. Prepared for the International Whaling Commission.

Kalland, A. (1993a). Management by totemization: Whale symbolism and the anti-whaling campaign. *Arctic*, *46*, 124–133.

Kalland, A. (1993b). Whale politics and green legitimacy: A critique of the anti-whaling campaign. *Anthropology Today*, *9*(6), 3–7.

Kedzlie, M. (2014). Why Does Japan Value Its Whaling Industry Over Its Reputation. *Think.* Accessed online https://think.iafor.org/whaling-in-japan/

Khoury, M. (2015). Whaling in circles: The makahs, The international whaling commission, and aboriginal subsistence whaling. *Hastings Law Journal*, *67*(1), 293–321.

Komatsu, M. (2005). *Yoku wakaru kujira yoron – Hogei no mirai wo hiraku [Understanding the whale controversy well – Opening up the future of whaling]*. Tokyo: Seizando.

Komatsu, M., & Takagi, Y. (2006). *Des baleines et des hommes [Whales and men]*. Tokyo: The Institute of Cetacean Research.

Krupnik, I., & Kan, S. (1993). Prehistoric Eskimo whaling in the Arctic: Slaughter of calves or fortuitous ecology? *Arctic Anthropology*, 30(1), 1–12. Accessed online http://www.jstor.org/stable/40316325

Lies, E. (2014, June 10). Japan shrugs off embarrassing court loss, vows resumption of Antarctic whaling. *Reuters*. Accessed online http://www.reuters.com/article/us-japan-whaling-idUSKBN0EL0YQ20140610

Marrero, M. E., & Thornton, S. (2011). Big fish: A brief history of whaling. *National Geographic Society*. Accessed online http://nationalgeographic.org/news/big-fish-history-whaling/

Morikawa, J. (2009). *Whaling in Japan: Power, politics and diplomacy*. Oxford: Oxford University Press.

Morishita, J. (2006). Multiple analysis of the whaling issue: Understanding the dispute by a matrix. *Marine Policy*, 30(6), 802–808.

Morishita, J. (2023). IWC68: Reflections on the future of the international whaling commission. *Whaling today*. Accessed online https://featured.japan-forward.com/whalingtoday/2023/01/13/iwc68-reflections-on-the-future-of-the-international-whaling-commission/

Papastavrou, V. (2006). In the name of science: A review of scientific whaling. IFAW. Accessed online http://ifaw-pantheon.s3.amazonaws.com/sites/default/files/legacy/In%20the%20Name%20of%20Science%20A%20Review%20of%20Scientific%20Whaling.pdf

Peterson, M. (1992). Whalers, cetologists, environmentalists, and the international management of whaling. *International Organization*, 46(1), 147–186.

Proulx, J. (1986). Whaling in the North Atlantic: From earliest times to the mid-19th century. Environment Canada, Parks Canada, National Historical Parks and Sites Branch, Ottawa.

Sekiguchi, T. (2007, November 20). Why Japan's whale hunt continues. *Time Magazine*. Accessed online https://content.time.com/time/world/article/0,8599,1686486,00.html

Sheehan, G. W. (1985). Whaling as an organizing focus in northwestern Alaskan eskimo society. In T. D. Price, & J. A. Brown (Eds.), *Prehistoric hunter-gatherers: The emergence of cultural complexity*. Orlando, FL: Academic Press.

Sommers, L. (1967). The Antarctic pelagic whaling crisis. *Tijdschrift Voor Economische En sociale Geografie*, 58, 126–134.

Surrency, E. (1964). International inspection in pelagic whaling. *International & Comparative Law Quarterly*, 13(2), 666–671.

Takahashi, J., Kalland, A., Moeran, B., & Bestor, T. C. (1989). Japanese whaling culture: Continuities and diversities. *Maritime Anthropological Studies*, 2(2), 105–133.

Tønnessen, J. N., & Johnsen, A. O. (1982). Translated by R. I. Christophersen. *The History of Modern Whaling*. Berkeley, CA and Los Angeles, CA: University of California Press.

Watanabe, H. (2009). *Japan's whaling: The politics of culture in historical perspective*. London: Trans Pacific Press.

Whiteman, H. (2015). Japan defies world as 'research' ship embarks on minke whale kill. CNN, December 1. Accessed online https://www.cnn.com/2015/11/30/asia/japan-whaling-research/index.html

3 Conceptual and methodological framework

The numbers and statistics concerning different aspects of whaling, including quantifiable data on whale populations, whaling quotas, etc. are being regularly collected and analyzed by several organizations, such as the Institute of Cetacean Research (ICR) in Japan, the intergovernmental North Atlantic Marine Mammal Commission (NAMMCO) and, of course, the International Whaling Commission. However, as the Chair of the IWC in 2016–2018 and the former Japan Commissioner to this organization Joji Morishita put it, commenting on the long-lasting stalemate at the heart of the whaling dilemma: "It is not about science, it is about feelings" (personal communication, November 2016). This means that the data on the health and abundance of whale stocks, no matter how reliable it is deemed and how meticulously and by which entity collected, has never been important enough to control the direction of the global whaling trends. Instead, other concerns, be it of moral rightness or political expediency, gained the dominant position.

After having conducted a preliminary study of the existing resources on whaling, my opinion was similar to what Dr. Morishita concisely expressed in the above phrase. The key to the long-lasting stalemate between anti- and pro-sustainable whaling forces lies in the *"feelings"* part of the story, or in *social meanings*, including people's emotional responses, perceptions, and contexts. It is on investigating these that this project focuses.

When it came to considering the methodology that would best fit the objectives of this research project, the rich contextual data that can be obtained through qualitative research tools appeared the most appropriate to answer the questions posed. The understanding that qualitative researchers seek through examining real-world situations as they naturally unfold (Golafshani, 2003), and the interpretive paradigm underlining this search for personal meanings are the foundations that were guiding this research.

The use of triangulation

The use of qualitative data is usually associated with questions of reliability and validity.[1] These issues were to a certain extent addressed through performing triangulation on two different levels: data triangulation and methodological triangulation. Moreover, following Olsen's argument, triangulation here is viewed

DOI: 10.4324/9781003255031-4

as an approach that can be used not merely for validation, but also for deepening and broadening the understanding of the research problem (2004). Such a notion of it is congruent with the overall concept of this research – it is exactly the lack of depth and breadth of the common perspective on whaling in Japan that motivated it. Thus, triangulation fits the idea of this study not only methodologically, but also conceptually.

Data triangulation

Data triangulation was achieved through the following steps:

1 collecting data from different sources – both secondary and primary data were used to arrive to the conclusions of this study – secondary sources included existing academic research on the subject, popular media pieces covering the subject – such as news and opinion articles, as well as broadcast media pieces (TV-shows and documentaries); primary data included the IWC archival materials and video recordings of the IWC meetings, video recordings of the press-conferences with the Japanese government and other Japanese whaling participants, as well as original research data in the form of interview recordings and fieldwork materials (photographs, video recordings, field notes);

2 collecting data from individuals belonging to different levels and types of engagement with the issue of whaling in Japan – including national government officials from the Whaling Affairs Office within the Fisheries Agency of Japan of the Ministry of Agriculture, Forestry and Fisheries; national government officials from the Ministry of Foreign Affairs; local governments' officials; representatives of the ICR; foreign PR consultants for the ICR; whaling-related business owners and workers; anti-whaling NGOs' representatives; pro-sustainable whaling NGOs' representatives; representatives of academia working on whaling-related topics; journalists regularly covering the topic; the Taiji Whale museum curator; fishermen in Taiji; policemen in Taiji; whalers dispatched from various previously whaling towns for research whaling; representatives of the delegations of member countries of the IWC at the organization's 67th meeting in Brazil in 2018;

3 collecting data from different locations – including the Japanese national government offices in Tokyo (Ministry of Agriculture, Forestry and Fisheries); the ICR, the Japan Whaling Association and Kyodo Senpaku Kaisha Ltd. whaling company offices in Tokyo; Kyodo Hanbai offices (a subsidiary of Kyodo Senpaku Kaisha Ltd.); offices of anti-whaling and pro-sustainable whaling NGOs in Tokyo; restaurants serving whale meat-based dishes in Tokyo; Tsukuji fish market in Tokyo; various whaling-related events' sites in Tokyo; the marine products section of Yokohama City Central Wholesale Market (Kanagawa Prefecture); the town of Wadaura (Chiba Prefecture) – Gaibo Hogei company office, the company's whale meat packing factory, the site where whales are dissected by the workers of Gaibo Hogei; the town of Taiji (Wakayama Prefecture) – local government offices, the office of Taiji Fisheries Association, the Taiji Whale Museum, other areas of the town associated with its whaling history and current whaling; whaling research site in Kushiro, Hokkaido;

4 collecting data during different times – two trips were made to both Wadaura and Taiji, for each locality one trip was planned during the non-whaling season, and another one during the whaling season; most of the research participants were met on a number of different occasions throughout the three years of active data collection process during 2016–2019 and I am still in regular communication with a few participants in 2023 as this book is undergoing final prepublication edits.

Methodological triangulation

It is not uncommon to combine several methods of data collection to reach one research goal. In the present research, however, not just methods, but two different but interrelated methodological approaches were used.

During the early stages of this project, the primary data gathering was imagined to be performed through in-depth semi-structured interviews based on a phenomenological approach only. However, while I was working on identifying interviewees and through them, connecting to others, elements of what ethnographers would call fieldwork were starting to emerge. It took several months of participating in various whaling-related events and gatherings for me to realize that this was not only a chance to meet the key figures in the Japanese whaling circles in order to interview them later, but also a valuable source of relevant data in itself – through observations of how people were showing up in these spaces and how they were interacting with each other and myself. This way, participatory observations and informal conversations were conceptualized as elements of fieldwork based on an ethnographic approach when I was already conducting them for some time. Additional data collection was then continued in that framework more proactively.

This is how the initial methodological ideas expanded in the course of this project, resulting in the application of two approaches – phenomenological and ethnographic – to the study of Japanese whaling and the people behind it. These two kinds of methodology can be considered complementary in terms of providing researchers with an opportunity to uncover deeper layers of a complex phenomenon while offering a perspective based on the perceptions and experiences of a group of participants (Maggs-Rapport, 2000). Thus, *methodological triangulation* was developed by combining phenomenological and ethnographic methodologies and conducting them in parallel, which enhanced the understanding of the Japanese whaling actors' positions and provided a more complete look at the issue of whaling from the Japanese perspective.

Phenomenology

"Phenomenology" can mean several things. It is considered to be both a philosophy (or a meta-theory, or a philosophical movement, or a disciplinary field in philosophy) and a comprehensive social science methodology. There is a certain ambiguity and uncertainty about the exact meaning and scope of it (Ehrich, 2005) and it continues to be open to revisions and new developments (Farina, 2014). It is widely used in a variety of disciplines. In this research project phenomenology was mostly applied as a method of data collection and analysis, but to give it some foundation and context, the philosophical underpinnings of this approach are also briefly discussed here.

Phenomenology as philosophy

The origins of phenomenology can be traced back to the late 19th century–beginning of the 20th century when positivism was the dominant philosophical view. The key figure who then challenged positivism and launched the phenomenological tradition was Edmund Husserl (1859–1938) – commonly considered to be the founding father of phenomenology. His work was developed by Martin Heidegger (1889–1976), Jean-Paul Sartre (1905–1980), and Maurice Merleau-Ponty (1908–1961). These four are the so-called Big Four of phenomenology. There are also other prominent figures in the field, including Hans-Georg Gadamer (1900–2002), Paul Ricoeur (1913–2005), and more recently and in a more practice-oriented way – Amedeo Giorgi (born 1931) and Max van Manen (born 1942).

Classical Husserlian phenomenology claims *centrality of experience from the conscious perspective of the subject*. Experiences can be of different types, including but not limited to ideas, perceptions, emotions, actions, social activities, etc. Experiences are characterized by "intentionality", which is a term Husserl adopted from one of his mentors – philosopher Franz Brentano. Initially, through this concept, Brentano was describing the human mental ability to refer to something that exists solely in one's mind, and he believed that the object of our *intentionality* (or our experience) is more of a mental product than a physical thing (Huemer, 2018). He also called it "reference to content" or the more well-known term of "mental or intentional inexistence" (Siewert, 2017). This idea was then further modified by him and his students, including Husserl, and in a simplified form it is now generally accepted that intentionality means that experiences are always directed at something in the world, they are *about* or *of* something (Smith, 2018). Intentionality comprises or sometimes can be equated to the mental representation of extra-mental objects (Lauwers, 2013; Siewert, 2017). Some also use intentionality in the sense of the total meaning of an experience, which always lies in the entirety of various perspectives on it (Ehrich, 2005). This concept was later added to the four crucial elements that characterize all the different types of phenomenology by Merleau-Ponty, along with reduction, description, and essences (Ehrich, 2003).

Husserl expressed the idea that seeing the essence of experiences is possible through the method of epoché, also called "bracketing" or "reduction". To be precise, according to Husserl there is not one reduction, but several different kinds, including epoché – the one that needs to precede others and is of a higher level. Epoché is an important element of the philosopher's phenomenological methodology and is based on Husserl's understanding that "objectivity itself is experienced objectivity" (Morley, 2010), but our "natural attitude" or unquestionable belief in the objective existence of the world around us (metaphysical realism) prevents us from seeing that and makes it impossible to truly unpack this. As Husserl thought, epoché involves setting aside the questions of reality of an object and its physical features and withholding judgment about the existence of the extra-mental world - then the experiences would appear in their original state – as they are given by one's consciousness (Farina, 2014).

The conceptualization of "bracketing" was continued by others. There were those who criticized it for its idealism and ignoring the factual impossibility of completely letting go of one's belief in the existence of the world around. Taking these criticisms into consideration Husserl himself moved to studying the environment that shapes experiences, or the *"lifeworld"* (Lauwers, 2013). However, in a more generalized way, seen as the idea of suspension of preexisting theories in order to be able to see and describe experiences or phenomena as they are, epoché was developed to become one of the foundations of phenomenological methodology as it is used now. Phenomenologists and social scientists in general use reductions when they "bracket" assumptions and presuppositions about the phenomena they study.

Morley (2010) maintains that the original Husserlian epoché is much deeper and more radical than what contemporary qualitative methodologies often take it for. In his opinion, it was not intended as a technique that can be simply applied to research in the form of recognizing one's assumptions before proceeding with data collection and analysis, but more of an existential "conversion experience" that takes profound knowledge of phenomenological writings and a lot of experience. It is telling that Husserl himself used the word "meditation" when talking about contemplative practices that constitute his phenomenological methodology (Morley, 2010). All while acknowledging Morley's judgment as valid and allowing for the possibility for some researchers to reach the state of epoché, for this research project the simpler, more practical applications of Husserl's ideas are accepted as useful and enriching the practice of qualitative inquiry.

Going back to the four key characteristics of phenomenology as listed by Merleau-Ponty, *essences* is the next important concept that is interrelated with the discussed above intentionality and reductions. It was also originally conceived by Husserl and later developed by his students and followers of his philosophy. Essences are understood as structures of the meaning of the phenomena in question, it is what distinguishes a particular phenomenon and makes it what it is (Dahlberg, 2006). According to phenomenology essences are not separable from the phenomena or experience. As Hill (2009) maintains, the question of what came before Husserl – essences or experience itself – is likely to never be answered, but will still be indefinitely debated by philosophers. Essences are what we automatically recognize just by glancing at objects. But as Husserl thought, our preconceptions and theories might lead us away from the essences and into their interpretations instead (Dahlberg, 2006). And here, we are again reminded of the concept of reduction, since it is through reduction, or "bracketing" our presuppositions, that we can reach out to the essences of phenomena.

The last of the four important elements of phenomenology according to Merleau-Ponty is *description*. Description of phenomena is the main method of phenomenological inquiry that is ultimately aimed at uncovering the essences of experiences.

Phenomenological descriptions should include variations - it is through the process of collecting and then analyzing the many variations of perspectives on experiences that one can penetrate the essence of the latter (Dahlberg, 2006). Although Husserl and the phenomenologists who followed his ideas have always called the concept "phenomenological description", if we take into consideration its end goal, it is, perhaps, more suitable to call it "phenomenological definition", as phenomenology endeavors to arrive at a generalized and, using Husserl's (Dahlberg, 2006) word, "invariant" meaning. Whatever the term, "phenomenological description" or "phenomenological definition" – capturing the essence is the goal, and that goal has successfully migrated from the original phenomenological philosophy to contemporary phenomenological applied research.

The methods of phenomenological description and reductions are not the only ones that are now being actively used in methodologies based on phenomenology. Other concepts developed by scholars who are well-known in this discipline are also prominently featured in social science research. Among the most important ones are Martin Heidegger's work on "Dasein" – on the subjects' exploring their "average everydayness"; and "being in the world" – on inseparability of subjects of experiences and their environment (Healy, 2011; Horrigan-Kelly, Millar, & Dowling, 2016). Heidegger is considered to be the founder of *hermeneutic* or *interpretive phenomenology* – aimed at understanding and interpreting the world (as opposed to describing it as in Husserl's phenomenology) as if it was a text. Heidegger argued that all description is already intrinsically interpretation (Ehrich, 2005).

In terms of bringing phenomenology in its original philosophical form closer to sociological methodological instrumentarium, the work of Alfred Schutz (1899–1959) stands out. In his principal work called "Phenomenology of the Social World," he focused on the meanings of social interactions and the structure of the social world among other themes (Barber, 2018). He developed the concept of "lifeworld" to mean the space where fellow human beings experience society and culture and act according to their influence (Goulding, 2005). Shutz also worked on the phenomenology of the political sphere, including issues of citizenship and racial equality (Barber, 2018).

Phenomenology as a qualitative research approach

As mentioned above, the first philosophers of phenomenology were not focused on the possibilities of using it for applied research. However, in the past four to five decades the underpinnings of Husserl and his disciples' thinking have been successfully employed in empirical studies. The phenomenological approach is commonly used in social science disciplines such as psychology, archaeology, sociology, gender and sexuality studies, etc. It has also gained popularity in applied sciences including but not limited to business, management, marketing, pedagogy medicine, and nursing.

It is important to note that when it comes to using phenomenology as a methodology in social science, arguably, there is no one agreed-upon procedure to follow – it is a complex landscape of preferred methods and focal points. Moreover, some argue that it is difficult to clearly differentiate research that belongs to phenomenology because of the various ways scholars treat this philosophy/methodology (Ehrich, 2005). Others add that it is more appropriate to talk about an approach and attitude toward participants and data than the methodology proper. The elements of phenomenological attitude include openness, humility, flexibility, and sensitivity (Lauwers, 2013).

Different disciplines and schools also use phenomenology in a number of diverse ways, and only some principal guidelines are commonly taken into consideration. To mention a few significant variances, it is useful to turn to the example of the two schools well-known for phenomenological research – the Utrecht School in the Netherlands (also referred to as the Dutch school) and the Duquesne School in the United States of America.

In hermeneutic (interpretive) phenomenology inspired by Heidegger's writings that was followed and further developed in the Utrecht school, and mostly applied to the field of pedagogy, the aim is to arrive at *insights* into participants' experience of a phenomenon. This is achieved through a wide variety of methods, the choice of which is strongly dependent on the circumstances of a concrete case. These may include researcher's own lived experience, interviews, observations, as well as descriptions of experiences in literature and art. The approach taken here is holistic – it offers rounded pictures of phenomena, and poetic – as the Dutch school researchers' value sensitivity and poetic language. Overall, the Dutch school can be characterized as creative and flexible. The Utrecht school has been criticized for never making clear "what phenomenological research was and what wasn't phenomenological research" (Lauwers, 2013), but the school is still considered one of the most influential ones and has continued to inspire researchers all over the world for more than five decades now (Lauwers, 2013).

On the other hand, in empirical phenomenological psychology, which is the focus of Duquesne School, it is considered that the main result of every study is an accurate *description* of all aspects of participants' experiences of a phenomenon. The researchers here follow a rather strict methodology and a predetermined process of analysis. The main source of data is the informant, and after data is collected, general statements reflecting the experience in question through its essential structures are produced.[2]

Whatever the approach there are elements of the phenomenological methodology that stay at the forefront of an inquiry of this kind. Typically, a researcher who adheres to this approach is interested in the *lived experiences* of their participants. Using the original phenomenological philosophical concepts, it is the participants' "natural attitude" (engaging with the well-known world in a relaxed state of mind not requiring any additional psychological or mental effort) when they are experiencing the "lifeworld" that phenomenological researchers inquire into (Bewan, 2004). Empirically this is grounded in the *meanings* those who are studied attach to the phenomenon in question. In other words, participants' subjective explanations of their own experiences are at the center of phenomenological research. The fact that these necessarily have to be lived experiences, not "second-hand" ones is of paramount importance. Otherwise, the data would appear as an interpretation of interpretation, which is counterproductive in the context of phenomenology. It is also useful to reiterate that the final goal of phenomenological research is not just a collection of insights/descriptions of experiences, it is a *description of the phenomena* that is the research problem through the descriptions of essences of experiences.

Another point to note is that phenomenological research is built upon *subjectivity*. If in other types of studies, subjectivity is seen as a problem that a researcher needs to account for, here it is an accepted part of the epistemological context. According to Levering (2006), subjectivity is the key to "granting personal meaning" and acknowledging that every person has the right to their own perspective. The scholar goes on to add that *intersubjectivity* also plays a significant role in human experiences, resulting in commonly held meanings. These are manifested in the rituals and language that representatives of the same social group share, and are also of interest to phenomenological researchers.

The acceptance of subjectivity leads us back to the perennial questions of truth and validity of data and findings based on data – how much trust should we put in a person's story about his own experiences and a researcher's interpretations of several personal experiential accounts. As is characteristic of most aspects of phenomenology, there is a lot of variety in the understanding of nuances of truth and what can be considered true within this discipline/methodology. The concept of truth has been featured in phenomenological writings throughout the history of their development – from Husserl to contemporary phenomenologists. It is telling that there is a full volume devoted to this problem that was published relatively recently – "Variations on Truth: Approaches in Contemporary Phenomenology" (2011) – where different theories of truth and parts of thereof are considered. Adding to the vigorous debates on this complex concept is outside the scope of this book. However, it is necessary to clarify a few aspects directly related to "truth" that were used to reinforce the overall conceptual foundation of this project.

The veracity of a particular story shared by a participant in this research project and interpretations of these were viewed through the prism of the *coherentist approach*. Although the coherence theory of truth is supported by some and refuted by others, in the context of phenomenology and this research its core suppositions provide appropriate guidance. Putting it simply, the coherence theory states that the acceptance of a proposition as true depends on whether or not we can logically and in an organized manner incorporate it into a system of other already accepted as true propositions related to the same topic. Talking about truth and interpretation in phenomenology Dahlstrom (2010) states that "the truth of interpretation is the coherence of its overriding meaning with the meanings of ... the things already interpreted". This understanding of truth as being founded in the coherence of interpretation is especially appropriate in situations when there is a lack of opportunity to verify a proposition against the facts of reality. One of the examples of such cases could be accounts of personal experiences – if we remember that emotions, thoughts, dreams., etc. belong to instances of experiences in phenomenology (along with actions). Although moral premises or feelings and emotions are considered non-truth-evaluable to begin with by some proponents of other truth theories,[3] some phenomenology-based views contend that *"true" is what people say is true when it comes to their own experiences* (Levering, 2006). Inherently there could be no one to know better about a person's feelings vis-à-vis a particular phenomenon than the person themselves. If there is no way to know better, and hence no way to oppose or confirm how one feels, it leaves us with no other option but to accept the person as the ultimate authority over their experiences and the statements about it as the ultimate truth in a given situation. This is not to say that no attempts at verifying participants' testimonials are made during a phenomenology-informed research study. Triangulation, previously discussed in this chapter, serves this purpose. The participants' non-verifiable statements can be reinforced by checking them against verifiable ones – for example, thoughts against actions. The final goal of getting to the essence of a phenomenon is achieved through composing the descriptions of a number of personal accounts as well as data from other sources (if the Dutch phenomenological school's methods are applied).

When the philosophical and conceptual foundations of phenomenology are taken into consideration, it is, perhaps, not surprising that one of the main instruments used in this type of inquiry is interviewing. Depending on the type of interviews, they can give direct and prolonged access to participants and permit the researcher to ask exploratory questions in search of participant-generated meanings. Other methods used in phenomenology include participatory observation, inspection of participants' diary entries, blogs, or vlogs, as well as notes by the researcher themselves (Lauwers, 2013). In this research project, *in-depth interviews* with the representatives of the Japanese whaling circles were used as the primary source of data for the part of the research inspired by phenomenology. In several cases, interview data was complemented by data from other sources, however, the interviews were viewed as the principal method.

In terms of the conduct of interviews there are but a few general guidelines from empirical phenomenologists. Bewan (2004) confirms that interviewing is the most widely used method in phenomenology, noting, however, that despite the popularity of the method, rarely any specific guidelines on conducting a phenomenological interview are given. Habitually, *purposive sampling* is used. The overall objective is to create a holistic picture of a phenomenon, hence, informants who have experiences related to the phenomenon of interest are targeted. As explained above, only those who have direct experience with the phenomenon are of primary interest to phenomenological researchers. Englander (2012) formulates the question phenomenologists have to ask themselves when identifying participants as "Do you have the experience that I am looking for?"

In qualitative research in general it is considered that the sample size can be relatively small, as a theme needs not to be recurrent to appear in the analysis and inform conclusions (Ritchie, Lewis, & Elam, 2003). In phenomenological interviewing in particular, given how much data is normally generated from a single interview, the sample size is also adequately small. No clear prescriptions are usually made about the necessary size of the group of participants a researcher should work within phenomenology. And, perhaps, none should be made. The guidance in terms of numbers is often sought through *saturation* - a stage in data collection when the same themes keep re-emerging and/or new ones do not emerge anymore (Saunders, Sim, & Kingstone, 2018).

In terms of interview questions methodological advice is also scarce but suitable given the variety of contexts phenomenology can be applied to. It is suggested that questions should be open-ended and broad to allow for the free expression of first-person perspectives on the part of research participants (Bewan, 2004).

For concrete methodological steps followed while inquiring into Japanese whaling-related individual's experiences of the phenomenon of Japanese whaling see Chapter 4.

Ethnography

Ethnography, being a method originally belonging to the cultural anthropological intellectual arsenal, at its core is an inquiry into the culture of a society, typically one on the small side – a social group. Culture in this type of study is understood in many different ways and often defined rather broadly. In this research culture is seen as shared by members of a social group values and beliefs, as well as acquired patterns of behavior, that can influence how individuals perceive their wellbeing, make decisions, and experience life in general (Marshall, 1990). Ethnography as a methodology is now used in a range of social science disciplines.

There are various types of ethnography, and depending on the purpose of research, both full and partial descriptions and analyses of groups are possible ways of conducting an ethnographic study (Goulding, 2005). In this research project, the main interest of investigation are the people who represent the many levels and ways of involvement in Japanese whaling. In other words, the one common thread identifying them as a group for the purpose of this study is their personal and/or professional connection to Japanese whaling, and it is this aspect of their life that is at the center of attention here. Ethnography presupposes the immersion of a researcher into their research field and ideally forming close relations with the participants (Dewan, 2018).

As a method of research for the particular case of Japanese whaling, ethnography was chosen first of all for its ontological and epistemological orientation. The ontological assumption of ethnography – the premise that the nature of what is studied varies depending on the situational and personal circumstances (Whitehead, 2014) – suits the idea behind this whole project that the Japanese people's understanding of the developments in whaling reveals dimensions of this issue that other groups do not appreciate. This project's epistemological standpoint is also aligned with the interpretive nature of ethnographic inquiry – intersubjectivity between the researcher and the researched is an accepted and, arguably, important part of it (Whitehead, 2014).

What is of most interest for this research is the so-called *emic* view – from inside the system. This was originally described in cultural anthropology as "from the native's point of view" (Malinovski, 1922 cited in Morris, Leung, Ames, & Lickel, 1999), and meant focusing on the one culture in question without outside references. Emic is the concept ethnographies are usually built upon, as it allows for themes and meanings previously unknown to the outsider – and the researcher is one such outsider in this case – to emerge from the participants and the field (Dewan, 2018). Gaining an emic perspective requires effort and a significant time commitment, as it is based on establishing and maintaining direct and continued contacts with the social group that is being studied, most often in the space they occupy – physical and/or virtual. That is why traditionally ethnography is considered to be a labor-intensive method of research (Goulding, 2015). Perhaps, the most indispensable and difficult-to-meet requirement for eliciting the emic of a social group one doesn't belong to, is the ability to suspend one's own worldview during ethnographic data collection and analysis.

In addition to the emic perspective, in this project the inquiry was taken further, and the *etic* perspective was also given attention. The etic is the view from outside, which includes comparing and contrasting several cultures (Olive, 2014) and links the findings from the studied field and social group to external factors, including economy, environment, political landscape, etc. (Morris et al., 1999). Researchers still argue about the necessity of integrating the emic and etic, dismissing one or the other based on the underlying conceptual and methodological differences (Morris et al., 1999). However, in the case of Japanese whaling, given its history and current developments, the external context could hardly be ignored. As will be demonstrated further, a significant portion of what was discovered in the field and discussed during both the informal conversations and the in-depth interviews with the participants, was to a great degree influenced by the Western perspective on whaling activities and their anti-whaling stance. The representatives of the Japanese whaling circles very often describe what they do and why they do it, what they feel, and how they perceive the problem in terms of how different it is from what the Western media, anti-whaling NGOs, and the anti-whaling "camp" in general tell about them. The decision to include the etic perspective to this project's methodological design was first informed by preliminary research, but later on, it was also supported by how the data collection was unfolding.

Another concern about claiming to focus solely on the emic insights was already hinted at here above and addresses the question of the researcher's subjectivity. Any social scientist comes into the field with their preconceived ideas, models of behavior, and beliefs. As much as it is desirable to distance oneself from these in order to take in the worldview of the "natives", it is, arguably, not entirely possible (Olive, 2014). During this project, a conscious effort to offset my personal biases on whaling was made, but it is important to note that I also believe that when you enter an unfamiliar environment, you are likely to bring the etic view along. This understanding served as further motivation to explore the hardly escapable researcher's own etic perspective on Japanese whaling and use it to enrich the understanding of the problem, rather than ignore it.

It was already mentioned that ethnography puts the researcher in the "field". Fieldwork is a crucial part of this method, or, how Agar (1980) puts it in his seminal book on conducting ethnographic research "The Professional Stranger", "the name for 'doing ethnography' is 'fieldwork'." Depending on the topic and goal of research fields can vary greatly, but they are all connected by the main idea that it is the natural environment of the studied people. It is their life (or certain parts of it), as it is. There are infinite possibilities in terms of who and what a researcher can investigate using ethnographic methodology, and there are infinite ways of how to do it. Real-life circumstances also inevitably influence how this methodology is applied, and from reading accounts of ethnographic studies by acclaimed social scientists, it becomes clear that the fieldwork part of a study rarely goes exactly as planned.[4] Agar (1980, p. 2) tells a funny story about a young graduate student who was given a task to research an Indian group without specifying where the group was located. The student goes to see the department head to ask where to start and how to do ethnography, but all she hears is a single piece of advice – to prepare a pencil and a notebook. After having started the fieldwork part of this project, I understood that this piece of folklore is only a slight exaggeration. Agar meant that no one can really train you to do fieldwork – your field is your best instructor and each step of your way is guided by the previous step. This was the case with this book. Reading extensively on both conducting research in the field and the Japanese people who participate in the country's whaling activities did help to prepare me for the real-life experiences, providing context and making data collection more efficient. However, it was the openness and generosity of the participants of this research, or in some other cases a lack of response and a cautious attitude from certain actors, that had the greatest influence on how this fieldwork eventually unfolded.

Notes

1 See, for example, Franklin, Cody and Ballan (2010), Golafshani (2003).
2 For more details on the two schools and the differences between their two distinctive approaches see Hein and Austin (2001) and Ehrich (2003, 2005).
3 For example, see the article of the Stanford Encyclopedia of Philosophy "The Correspondence Theory of Truth" (retrieved from https://plato.stanford.edu/entries/truth-correspondence/)
4 There is a vast array of articles and books that support this statement. For example, the already mentioned "The Professional Stranger" (Agar, 1980). Another captivating account – of a six-year-long project by Adler (1993) – "Wheeling and Dealing: An Ethnography of an Upper-level Drug Dealing and Smuggling Community" – also clearly illustrates this point. "At Home in the Street: Street Children of Northeast Brazil" by Hecht (1998) can also be considered an example of going where the field takes you rather than relying on a rigid plan.

References

Agar, M. (1980). *The professional stranger: An informal introduction to ethnography*. Cambridge: Academic Press.
Barber, M. 2018. Alfred Schutz. In E. Zalta (Ed.), *The Stanford encyclopedia of philosophy*. Retrieved from https://plato.stanford.edu/archives/spr2018/entries/schutz/
Bewan, M. (2004). A method of phenomenological interviewing. *Qualitative Health Research, 24*(1), 136–144.
Dahlberg, K. (2006). The essence of essences: The search for meaning structures in phenomenological analysis of lifeworld phenomena. *International Journal of Qualitative Studies on Health and Well-Being, 1*, 11–19.

Dahlstrom, D. Truth and interpretation. In P. Vandevelde and K. Hermburg (Eds.), *Variations on truth: Approaches in contemporary phenomenology* (pp. 209–224). New York, NY: Continuum.

Dewan, M. (2018). Understanding ethnography: An 'Exotic' ethnographer's perspective. In P. Mura, & C. Khoo-Lattimore (Eds.), *Asian Qualitative research in tourism. Perspectives on Asian tourism* (pp. 185–203). Singapore: Springer.

Ehrich, L. (2003). Phenomenology: The quest for meaning. In T. O'Donoghue, & K. Punch (Eds.), *Qualitative educational research in action doing and reflecting*. New York, NY: RoutledgeFalmer.

Ehrich, L. (2005). Revisiting phenomenology: Its potential for management research. In *Proceedings challenges or organisations in global markets, British Academy of Management Conference* (pp. 1–13). Oxford: Said Business School, Oxford University. Accessed online https://eprints.qut.edu.au/2893/1/2893.pdf

Englander, M. (2012). The interview: Data collection in descriptive phenomenological human scientific research. *Journal of Phenomenological Psychology*, *43*, 13–35.

Farina, G. (2014). Some reflections on the phenomenological method. *Dialogues in Philosophy, Mental and Neuro Sciences*, *7*(2), 50–62.

Franklin, C., Cody, P., & Ballan, M. (2010). Reliability and validity in qualitative research. In B. Thyer (Ed.), *The handbook of social work research methods* (pp. 335–374). Thousand Oaks, CA: Sage.

Golafshani, N. (2003). Understanding reliability and validity in qualitative research. *The Qualitative Report*, *8*(4), 597–606.

Goulding, C. (2015). Grounded theory, ethnography and phenomenology: A comparative analysis of three qualitative strategies for marketing research. *European Journal of Marketing*, *39*(3/4), 294–308.

Healy, M. (2011). 'Heidegger's contribution to hermeneutic phenomenological research. In G. Thompson, F. Dykes, & S. Downe (Eds.), *Qualitative research in midwifery and childbirth: Phenomenological approaches* (pp. 215–232). London, England: Routledge.

Hein, S., & Austin, W. (2001). Empirical and hermeneutic approaches to phenomenological research in psychology: A comparison. *Psychological Methods*, *6*(1), 3–17.

Hill, C. (2009). Husserl and phenomenology, experience and essence. In A.-T. Tymieniecka (Ed.), *Phenomenology and existentialism*. Dordrecht, The Netherlands: Springer.

Horrigan-Kelly, M., Millar, M., & Dowling, M. (2016). Understanding the key tenets of Heidegger's philosophy for interpretive phenomenological research. *International Journal of Qualitative Methods*, *15*(1). Accessed online https://doi.org/10.1177/1609406916680634

Huemer, W. (2018). Franz Brentano. In E. Zalta (Ed.), *The stanford encyclopedia of philosophy*. Accessed online https://plato.stanford.edu/archives/fall2018/entries/brentano/

Lauwers, H. (2013). Phenomenological research in educational sciences or learning how to cope with ambiguity. Onderzoekscentrum Kind & Samenleving (Childhood & Society Research Centre). Accessed online https://k-s.be/medialibrary/purl/nl/0044597/phenomenological%20research.pdf?download=true

Levering, B. (2006). Epistemological issues in phenomenological research: How authoritative are people's accounts of their own perceptions? *Journal of Philosophy of Education*, *40*(4), 451–462.

Maggs-Rapport, F. (2000). Combining methodological approaches in research: Ethnography and interpretive phenomenology. *Journal of Advanced Nursing*, *31*(1), 219–225.

Marshall, P. (1990). Cultural influences on perceived quality of life. *Seminars in Oncology Nursing*, *6*(4), 278–284.

Morley, J. (2010). It's always about the epoché. In T. Cloonan (Ed.), *The redirection of psychology: Essays in honor of amedeo giorgi* (pp. 293–305). Quebec City, Canada: University of Quebec Press.

Morris, M., Leung, K., Ames, D. R., & Lickel, B. (1999). Views from inside and outside: Integrating emic and etic insights about culture and justice judgment. *The Academy of Management Review*, *24*(4), 781–796.

Olive, J. (2014). Reflecting on the tensions between emic and etic perspectives in life history research: Lessons learned. *Forum Qualitative Sozialforschung / Forum: Qualitative Social Research*, *15*(2). Accessed online http://www.qualitative-research.net/index.php/fqs/article/view/2072/3656#g31

Olsen, W. (2004). Triangulation in social research: Qualitative and quantitative methods can really be mixed. *Developments in Sociology*, *20*, 103–118.

Ritchie, J., Lewis, J., & Elam, G. (2003). Designing and selecting samples. In J. Ritchie, & J. Lewis (Eds.), *Qualitative research practice: A guide for social science students and researchers* (pp. 77–108). London: Sage.

Saunders, B., Sim, J., & Kingstone, T., Baker, S., Waterfield, J., Bartlam, B. et al. (2018). Saturation in qualitative research: Exploring its conceptualization and operationalization. *Quality and Quantity*, 52(4), 1893–1907.

Siewert, C. (2017), Consciousness and intentionality. In E. Zalta (Ed.), *The stanford encyclopedia of philosophy*. Accessed online https://plato.stanford.edu/archives/spr2017/entries/consciousness-intentionality/

Smith, D. (2018). Phenomenology. In E. Zalta (Ed.), *The stanford encyclopedia of philosophy*. Accessed online https://plato.stanford.edu/archives/sum2018/entries/phenomenology/

Whitehead, T. (2014). What is ethnography? Methodological, ontological and epistemological attributes. *Cultural Ecology of Health and Change. Ethnographically Informed Community and Cultural Assessment Research Systems (EICCARS) Working Paper Series*. Accessed online https://static1.squarespace.com/static/542d69f6e4b0a8f6e9b48384/t/57951c7815d5dbb62e94c177/1469389948576/Intro%2Bto%2BCEHC.pdf

4 Conducting interviews and fieldwork in the context of Japanese whaling

In this research project, phenomenology and ethnography were not used consecutively with separate data sets as a result, but in combination, generating a mixture of data that was then analyzed as a whole, with the development of relationships between different pieces of data that helped form a rounded picture of the Japanese whaling landscape. Such convergence of data from different methodological origins is known as the "inter-method" (or "between-method") approach, and combining phenomenology and ethnography is considered well-suited to respond to certain research goals.[1]

In-depth interviews

As discussed in the previous chapter, interviewing was chosen to be one of the central methods of data collection for this study. Interviewing was aimed at:

1 Inquiring into participants' experiences of the phenomenon of Japanese whaling in the context of the changing status of this activity in the country and abroad;
2 inquiring into how the participants understand and account for their routine activities in relation to the dynamics of whaling in Japan;
3 inquiring into the features characterizing Japanese whaling-related people as a group, for example, intragroup communication;
4 and more generally, obtaining information that could fill in the gaps in the understanding of the issue of Japanese whaling.

In-depth interview respondents

The most important criterion for qualifying to be a participant in this study was for a person to be directly involved in Japanese whaling on some level and in some capacity. Active past involvement was also considered to fulfill this requirement. However, it is not always as easy as it might first appear to select the right participants, even when there is only one key criterion. The researcher, who in many cases does not belong to the environment they intend to study, might not know who experiences the phenomenon in question, who belongs to the social group of interest, or where to find them. This was the case with this project. To resolve

DOI: 10.4324/9781003255031-5

this problem, at first, I consulted popular media sources and academic literature to identify potential *gatekeepers* – organizations and/or people who are able to provide access to prospective participants. The first direct contact was made with the Institute of Cetacean Research (ICR) and their person of contact for such inquiries. From there, introductions to other actors followed, including representatives of the Japan Whaling Association (JWA), representatives on the ministerial level (Japanese Ministry of Agriculture, Forestry and Fisheries (MAFF) and the Ministry of Foreign Affairs), as well as industry professionals. Those who became study participants also often suggested to talk to one or two of their colleagues or advised to reach out to this and that person for further connections and events' information. Thus, *the snowball method* – when the sample expands through participants' recommendations (Groenewald, 2004), was used.

The number of people participating in phenomenological studies as well as ethnographic ones varies in accordance with specific situations and the research goal, as discussed in the previous chapter. Phenomenology generally considers up to 10 people as a sufficient number to reach saturation when conducting in-depth interviews (Groenewald, 2004), while in actual studies, the number commonly varies from as few as one participant to as many as 20. In ethnography, a larger number of participants is common, but it also heavily depends on the study design. In this project, the list of in-depth interview participants included 44 people. With the help of context from fieldwork, this number of interviewees allowed for data saturation.

Data collection through in-depth interviews

As prescribed by both phenomenology and ethnography, the questions asked during the in-depth interviews were *open-ended and exploratory*. The objective was to get the participants comfortable to talk about what they imagined their experiences to be like and what they believed was most important about the lifeworld of whaling, transferring the agency to the representatives of the Japanese whaling community.

Since the pool of participants included people with varying degrees of willingness to open up to an "outsider" about their views on the topic many of them were perceived as sensitive, and no one interview was like the other in terms of the time it took, level of involvement of the participant, and the quality of data obtained. Most interviews lasted *1–2 hours*. Exceptions included several interviews that were as short as 40 minutes and one that took as long as 6 hours. Interestingly, the ones on the shorter side of the spectrum were with people whose experiences were initially identified as the least discussed in the published research and the popular media accounts of whaling – the people who work with whaling directly, so to speak on the ground (or in this case at sea), including fishermen, harpooners, and specialists in the dissection of whales' bodies. They were also the most difficult to reach out to, some were reluctant to participate, and some of them opted out of having their interviews recorded. That is one important reason why it was appropriate to also use methods other than in-depth interviewing to reach the goals of this project – fieldwork gave a chance to observe and better connect to the people who have important perspectives on the issue of Japanese whaling, but were not readily willing to share their thoughts.

In-depth interviews' data analysis

The data collected through interviews was analyzed using *the content analysis* technique. *Relational analysis* when concepts and themes are not only found and described, but additionally relationships between them are investigated was also employed. This latter technique is especially useful when merging data generated using different methodological approaches (Robinson, 2011), which was the case with this research project.

Fieldwork

As mentioned above, the course of fieldwork conducted for this research was guided by the field itself and the participants of this research. The very decision to engage in fieldwork was taken when the fieldwork had already effectively started. It was only then that this crucial part of this project was properly conceptualized and continued with more intent and by the adequate methodological prescriptions.

Fieldwork was deemed necessary for this research, as 1) people do not always act the way they say they act (Agar, 1980); 2) it is better to see something once than to hear about it a thousand times (a proverb with equivalents in several different cultures/languages). The first statement is an important premise of social studies – people do not always know how to express what they experience, and at times they do not think things are worthy of sharing. In such cases observing participants' activities gives more details and provides for a better understanding of the experiences described during interviews. The second statement is more than illustrative of this study. I would have never been able to imagine what it is like to be a whale dissection specialist should there be no chance to observe this process in person; and no matter how many times you hear that "a whale can feed half a village" (personal communication, 2018), I had no point of reference before I saw a Baird's beaked whale's body being dissected in Minamiboso – and although it is far from being the biggest whale species, its size at close to 11 meters long was nothing short of impressive and certainly looking like it could provide food for a big group of people.

In the course of this research the following fieldwork data collection strategies were used:

- Direct observation – without participating in researched subgroups' official activities;
- participant observation – with (partial) participation in researched subgroups' official activities;
- informal one-on-one and collective discussions during researched subgroups' official activities;
- informal ethnographic interviews;

- participation and discussions in subgroups' informal gatherings;
- video footage provided by Women's Forum for Fish (WFF);
- consulting printed materials provided by the local government office of Taiji, the Japan Whaling Association, and the Institute of Cetacean Research;
- self-reflection.

The fieldwork included several field trips to the whaling towns of Minami-boso (Chiba Prefecture); Taiji (Wakayama Prefecture); and Kushiro (Hokkaido Prefecture). Not least importantly, attending and participating in numerous events in the Tokyo and Yokohama area that promoted whaling culture proved to be very informative for this research. Most of these were organized by the Whaling Affairs Office within the Fisheries Agency at the Ministry of Agriculture, Forestry and Fisheries of Japan (MAFF), as well as the Japan Whaling Association and an NGO Women's Forum for Fish.

The events I had the privilege to be invited to – their very existence, the regularity with which they are held, the effort put into their organization, and the way they were unfolding – became a valuable source of data. The opportunity to not only conduct formal interviews but to be present in several different whaling-related spaces, participate in informal conversations, and observe – it all has added depth to this study. It was a unique chance to observe the Japanese whaling-related people and their "natural" environment and also gave an opportunity to meet with a lot of the participants not on one, but on several occasions. This helped to refine, contextualize, and verify the data collected during the in-depth interviews.

It is generally agreed upon that in social sciences fieldwork is *prolonged contact* with "natives" and their environment (Goulding, 2005). Exactly how long depends on the research design. It could be days, months or years. In addition to the need to meet research objectives, major factors determining the time frame could be a preexisting familiarity with the field, as well as the level of experience of the ethnographer.

In the case of this research project, I was not previously close to anyone or any organization/institution in Japan that was engaged in whaling on a political or practical level. Additionally, research participants, although all related to Japanese whaling in one way or another, belonged to several subgroups each representing a different way or level of engagement with the issue of whaling. These sub-groups had different primary work locations, and the non-regular activities some of them organized and participated in (events, workshops, meetings, etc.) were occurring at significant intervals of several weeks and in a lot of cases several months. These were some of the reasons I stayed in the field for more than two years physically and maintained a virtual presence for several years after.

Some of the most important happenings in the whaling world globally are the International Whaling Commission Committee Meetings. Since the 1970s, the Japanese delegation was one of the most pro-active ones during these, and a lot of the people that were interviewed or encountered during my fieldwork were usually either present at the IWC Committee meetings or at the very least knew a lot about the proceedings. These meetings receive often biased and/or partial coverage in the media, the quality of streaming the organization provides during the event's sessions leaves much to be desired, and, as is usually the case at big-scale events, the communication that takes place on the way to sessions, during breaks, and at receptions is at times just as important as the official part. That is why being an attendee of one of the IWC Committee Meetings was important for this project. That prolonged the fieldwork since from 2012 the IWC decided to hold their Committee Meetings biennially – meaning that there is only one chance in two years to observe one in person.

Limitations

As the participants were not compensated for their time, it was most appropriate to prioritize their preferences for the time and place of the interviews. The study was also overall conducted without using any specially regulated and equipped research facilities. These factors resulted in variations, in certain cases significant, in the interviews settings.

It is necessary to reiterate here, that, as explained in the note on language below, two out of three parts of the methodology of this research heavily rely on my high level of both spoken and written Japanese language comprehension. Although I hold the highest level of the Japanese Language Proficiency Test – N1 – and have experience living in Japan and working in Japanese language environments for over a decade, I am not a native Japanese speaker. Some of the interviewees were using dialectisms and professionalisms related to the whaling and/or fishing industries, and some were speaking fast. Efforts were made to minimize the possibility of the details of the data getting lost in translation. If things were not completely clear during the data collection stage, clarifications were asked from the study participants whenever possible. When the data was being analyzed, the help of a native speaker was sought to review unclear passages in the audio- or video- recordings. However, while conducting the interviews and especially while doing fieldwork, time and circumstances do not always allow for lengthy explanations. In these instances, I could only rely on my on-the-spot Japanese communication skills.

These factors were offset as much as possible by the described in the above sections data and methodological triangulation.

A note on language

Although language is undeniably the principal tool of collecting ethnographic data, and in a lot of cases ethnographers have to switch between two or more languages while in the field and during interpretation and analysis, very few ethnographers address the issue of language (in)competence in their writing (Borchgrevink (2003); Gibb and Iglesias (2017).

I was very often asked the question about language when I shared the research idea during the early stages of this project's planning or any of the details of fieldwork or interviews when it was already mid-way. Was Japanese or English used for interviews? How about the events that I attended and participated in? How good does your Japanese have to be to be able to tease out underlying meanings from the data? These and many other language-related questions were raised in numerous conversations, and at times I was asking the same questions from myself. Professional interest from colleagues doing research in similar fields, as well as casual curiosity from friends and acquaintances contributed to my decision to address this question in writing here.

The *de facto* national and most widely spoken language of Japan is Japanese.[2] It has numerous dialects, the Tokyo dialect being the standard one. Although English is commonly taught in schools, not many Japanese people feel entirely comfortable communicating in English. The reasons for this are debated. Some of the explanations offered so far include issues with foreign language education, mainly criticized for its focus on translation rather than communication (Mitchell, 2017) and insufficient number of hours dedicated to language learning in schools (Tsubota-Newell, 2017); the so-called pronunciation anxiety (Lund, 2015) and Japanese people lacking motivation to learn English, as they see no need to use it in their own country (Morita, 2017). The Japanese Ministry of Education, Culture, Sports, Science and Technology has been introducing various policies to improve the relatively poor scores of Japan in international English language tests, however, the effectiveness of these policies is questioned (Reesor, 2003) and the results are yet to be seen.

The fact that Japanese people rely predominantly on their native language for both personal and professional communication is important for how the methodology of this project was designed. Nearly all written and oral communication had to occur in Japanese. This included reaching out to potential interviewees via email and phone, conducting the interviews themselves, the logistical organization of all elements of fieldwork, participating in Japanese whaling-related events, and communicating with the "natives" on the spot. There were only a few exceptions to this – the non-Japanese interviewees all preferred to use English. Thus, conducting this research using the chosen methodology required a high level of Japanese language proficiency.

The second point related to language is important methodologically and also contributes an insight relevant to this study's research objectives. Ethnographers generally agree that an operational familiarity and even better good command of the language native to the field of interest is a desirable skill (Gibb & Iglesias, 2017). Indeed, my ability to communicate in Japanese has been extremely important for the successful completion of this research. It facilitated the collection, interpretation, and analysis of the original data. Perhaps more importantly, it helped to gain the trust of this study's participants. The initial mistrust was expected given the many cases of failed attempts to communicate their feelings to international journalists and activists. But gradually people seemed to be getting more comfortable around me. It was definitely the function of how much time I spent in the field – showing up again and again tends to persuade people that you are serious about your work. But the ease with the "natives" native language also gives people reassurance that you are genuine in your intentions to see their side of the issue.

Language proficiency aside, what also helped to establish strong connections and reveal some of the most interesting insights of this research was active and engaged listening. One of the motivating forces behind this study and one of its conclusions is that the Japanese side of the story as seen by the very people who are its main characters has not been previously given enough attention. Yes, international outcry about the Japanese whaling has been there for a while, but have we been really listening to their response, or just talking to ourselves?

Notes

1 See, for example, Maggs-Rapport (2000).
2 There is no law defining Japanese as the official language of the country. See the Legislative Bureau House of Councilors quoting Sato Haruo (2006) on this here http://houseikyoku.sangiin.go.jp/column/column068.htm (in Japanese).

References

Agar, M. (1980). *The professional stranger: An informal introduction to ethnography.* Cambridge: Academic Press.

Borchgrevink, A. (2003). Silencing language: Of anthropologists and interpreters. *Ethnography, 4*(1), 95–121.

Gibb, R., & Iglesias, J. (2017). Breaking the silence (again): On language learning and levels of fluency in ethnographic research. *The Sociological Review, 65*(1), 134–149.

Goulding, C. (2005). Grounded theory, ethnography and phenomenology: A comparative analysis of three qualitative strategies for marketing research. *European Journal of Marketing, 39*(3/4), 294–308.

Groenewald, T. (2004). A phenomenological research design illustrated. International Journal of Qualitative Methods, 3(1), article 4. Accessed online http://www.ualberta.ca/~iiqm/backissues/3_1/pdf/groenewald.pdf

Lund, E. (2015, April 10). Pronunciation anxiety: Many Japanese people don't want to speak English unless it's perfect. *Japan Today.* Accessed online https://japantoday.com/category/features/lifestyle/pronunciation-anxiety-many-japanese-people-dont-want-to-speak-english-unless-its-perfect

Mitchell, C. (2017). Language education pressures in Japanese high schools. *Shiken, 21*(1), 1–11.

Morita, L. (2017). Why Japan needs English. *Cogent Social Sciences, 3*(1). Accessed online https://www.cogentoa.com/article/10.1080/23311886.2017.1399783

Reesor, M. (2003). Japanese attitudes to English: Towards an explanation of poor performance. *NUCB JLCC, 5*(2), 57–65.

Robinson, O. (2011). Relational analysis: An add-on technique for aiding data integration in qualitative research. *Qualitative Research in Psychology, 8*(2), 197–209.

Tsubota-Newell, I. (2017, October 29). Why do Japanese people have trouble learning English? *The Japan Times.* Accessed online https://www.japantimes.co.jp/opinion/2017/10/29/commentary/japan-commentary/japanese-trouble-learning-english/

5   The pillars of modern-day Japanese whaling

Japanese whaling is an example of a problem that demonstrates the interconnectedness of the modern-day world. If I were to examine Japanese whaling a mere 150 years ago, which is short in terms of human history, there would be much less mention of countries other than Japan itself. Japan adopted modern Norwegian-style whaling at the beginning of the 20th century, and this has had a profound effect on the development of the industry and the discourse around it in the long run, both in Japan and around the world. Nowadays it is virtually impossible to talk about whaling in one given country, without mentioning many others. Although the conceptual and methodological framework of this research project presupposed close attention first and foremost to the subject of this study, when conducting the desk research, as well as the fieldwork and interviews for this project, "the West", "anti-whalers", "foreign activists", "foreigners", and "foreign researchers" kept coming up time and again. This chapter shows that Japanese whaling today opposes Western preconceptions and misconceptions, but is also shaped by the latter, both in actions and in perceptions of itself. The first section of the chapter directs the reader toward formulating the right questions in the context of Japanese whaling, and the sections following it present the interconnected aspects of Japanese whaling that in the course of this research were identified as its "pillars" or the concepts and ideas that can guide one through the often complicated and conflicting information on the problem.

On the imagined "super-whale" and asking the right questions about Japanese whaling

The main premise of this study is that the Japanese side of the whaling debate is not adequately represented in the Western media and, subsequently the Western people's perceptions of it are wrong. It is not by accident that I use the word "wrong" so bluntly here. In this case, its use is warranted in the sense that what is not completely right or correct is, logically speaking, wrong. Not knowing the full picture, but proceeding to form an opinion will often bring one to the wrong position. And this is exactly the situation observable with Japanese whaling. Lack of attention to details and shallow understanding of the Japanese stance are at the core of the Western reactions to this issue.

DOI: 10.4324/9781003255031-6

It needs to be reiterated here that this book does not intend to deny the potency of animal rights-based perspectives, or non-human ecocentric attitudes to the environment. These concepts have been continuously gaining both theoretical and practical strength, and it is most interesting to see how their wider acceptance and application will change the world in the future. The analysis of their viability for all food systems currently existing in different areas of the world is outside of the scope of this book, but it needs to be mentioned here that their full adoption will require more political willpower and technological advances than we are witnessing in 2023.

Of more relevance to this project is the fact that the general attitudes of the West toward Japanese whaling rarely represent people who embrace comprehensive and well-thought through ecocentric views and put them into practice. In most cases, the accusations of countries and people who whale, and, as The Guardian put it in one of its articles "the international condemnation" of Japan (McCurry & Weaver, 2018), are based on attributing exclusive features to all whales without any scientific basis, as well as a lack of knowledge on the current state of the various whale species and whaling techniques.

The word "wrong" used above mirrors just that lack of subtlety with which the anti-whaling perspectives are treating Japanese whaling. The international community, or at least the most vocal part of it, has labeled Japanese whaling a "problem" that needs to be solved by elimination. Only then will the good triumph over the bad. No compromises can be accommodated, and no factual negotiations are necessary.[1] The West just wants to save the "majestic" creatures[2] from the "cruel, archaic and unnecessary industry".[3] This cavalier formulation of what constitutes the problem does little but promote populist rhetoric in its sensationalist form. (Ab) using anti-whaling to create and reinforce a "green" image costs most governments nothing – their own whaling industries are long obsolete, but it brings some important dividends making them look concerned with environmental protection.[4] Similarly, even if some countries' representatives sympathize with Japan's arguments, it would be difficult to backtrack now that the popular opinion in their communities skews strongly anti-whaling.

The media, as well as academic publications on Japanese whaling have plenty of examples of authors being puzzled by the *why* question – this was already brought up in Chapter 2. *Why* does Japan do what it does, *why* is it not receptive to our (Western, developed, English-language) pleas for ending the heinous practice of whaling, *why* does it defy the rest of the world[5] when it does not usually do that in other areas of international environmental concern, *why* does it ignore the damage whaling does to its reputation. Most recently, the *why* did Japan go as far as leaving the International Whaling Commission (IWC) question was added to the list. While wondering *why* is in itself not problematic and is a major driver of progress, in the issue of whaling a lot of these questions are premature at best and arrogant at worst. Many of them come from a self-righteous perspective that presupposes that one knows with certainty what is right and wrong, and those right and wrong are constants with no time-, space-, actors-, and background-dependent variability. Many of the publications and comments that have the *why* question at their core also imply or directly state that everything has been done to fix the Japanese stubbornness – we have threatened them with trade sanctions,[6] we have shamed them with demonstrations,[7] we have pursued them into the Antarctic Ocean every summer for 7 years, where we attacked them and showed that in detail on TV,[8] we have gone as far as personal attacks on people who support Japanese whaling.[9] In that light the Japanese actions become even more puzzling and the reason for them to not halt the hunts escapes the anti-whalers' minds yet again.

Seeking answers to the *whys* and through them trying to find keys to make the Japanese stop whaling form the basis of a problematic and fixative approach to this issue. Looking at what kind of argumentation usually follows the indignant accusations of Japanese whaling, it becomes clear that what we often observe is ignorance among those who have no time or interest to care for details. Among the political elites and scientists at the international fora such as the IWC, there are many of those who are aware of the specifics of the problem and of the relative soundness of both the Japanese overall stance and the proposals they were putting forward. This became evident while I was attending the 67th IWC Committee meeting that took place in Florianopolis, Brazil in September 2018. Quoting Robert Suydam, who is a wildlife biologist and the chair of the IWC Scientific Committee at that time:

> It is really clear that there are whale stocks that are doing well, and in some cases doing really well. And they actually can sustain a hunt and really the IWC agrees with that. [IWC approved quotas for some aboriginal hunts], but it is true for commercial as well.
>
> (personal communication, September 14, 2019)

However, since the constituents or the audience of the delegations to the IWC back in their home countries, surrounded by a wall of media sensationalism and misinformation on whaling, are not ready (and since the state of affairs has not changed for the past several decades, they will hardly be ready in the foreseeable future) to accept any compromises and concessions to the pro-sustainable whaling forces, it is easier and more logical for the traditionally anti-whaling countries to keep to their regular "no commercial whaling at all in no form and quantity under any circumstances" standpoint.

The same logic oftentimes guides publishing houses, as well as scientific and popular journals' editorial boards. The overpowering popular image of noble activists and the fragile prize they are fighting for – the life of the super-whale[10] – makes it difficult to incorporate conflicting elements into the whaling discourse. Publishing a paper or commentary that speaks in favor of sustainable whaling could be a risk that not many are prepared to take because they have nothing to gain from it, but they could potentially lose in readership, sponsorship or suffer other reputational repercussions.

As an example of this, one participant of this study, being a marine scientist and working with samples obtained from the Japanese scientific whaling programs, shared his experience of having difficulties with publishing the results of his and his colleagues' work in academic journals. He attributed this to the reputation-related reasons described above (personal communication, September 13, 2017). Indeed, Japanese scientific whaling was one of the most criticized aspects of the country's whaling policies that was prominently condemned in the media in various visual and textual formats. Hence, the results of these studies would rarely find their way to the broader scientific community.

This situation creates what can be metaphorically compared to a black hole that absorbs and erases all the bits of information that counter the predominant anti-whaling discourse. It is not that the information is not available, if one does their own research. It is just more often than not lost in a powerful flow of misconceptions, generalizations, and myths about whales and whaling, as well as labels that demonize and mischaracterize whalers and whaling-related people.

That is why it took me several years of fieldwork to be able to claim some understanding of the problem. Most of the 44 interviewees and other participants of this research as per its design were people who have been in one way or another involved in the issue of whaling for long periods of time – some for 20 years and longer. They all possess intimate knowledge of the industry and a deep understanding of the problem, which they were generous enough to share with me in the framework of this project. Without a strong incentive to do meticulous research and with no opportunity to reach out for in-person expert opinion, it is not easy to find visual and textual resources that contradict the mainstream ideas about whaling in a logical, comprehensive, and accessible manner.

Advice from participants and hours of searching did make some thought-provoking perspectives on whaling surface online. There are documentary films that include the perspectives of people from small whaling towns in their narratives;[11] recordings of interviews and talks where people express more nuanced views on whales and whaling;[12] blog posts dedicated to explaining some specific aspects of the whaling controversy;[13] books that inquire into cultural value of whaling for local communities;[14] and when it comes to primary data there is also the open-access collection of all the documents of the IWC. These do not necessarily express the most radical pro-whaling ideas and are not all made by the most committed defenders of whale meat consumption. They merely allow for the possibility of a more complex story about whalers and whales. They also do not (only) ask the *whys* about Japanese whaling and do not directly seek solutions to the often vitriolic debate surrounding it – they just explore and add details to *what* is happening. And it is this small pool of *what*-based materials that this study was drawing upon and also intends to contribute to in its own way.

What as a starting point of inquiry has been overlooked within the issue of whaling. The *whys* are prominent, but the *whats* are hidden, regardless of the fact that many of the latter are left unanswered in the prevalent whaling discourse. This dubious tendency prevents the public from forming their own balanced opinion about Japanese whaling and undermines the possibility of a productive exchange of views on the matter.

Let's take a very basic question that should be one of the first to come to mind: *What* species of whales are the Japanese targeting? There are more than 80 species of cetaceans that include the informal grouping of "whales" along with dolphins and porpoises. Not all of them are endangered – it is a scientific fact that has been long-proven (see Robert Suydam's remark quoted above in this section), and some of them have arguably never been at risk, even at the point in time when a blanket moratorium on commercial whaling was introduced in 1986. This latter group includes common minke whales that were one of the key interests of the Japanese whaling industry prior to 1986 and have risen to that rank again once the country withdrew from the IWC. This seemingly straightforward question about species, based on common-sense knowledge that there is definitely more than one type of whale (off the top of their head any reader would probably be able to name at least the blue, the humpback, and the orca) is interestingly very rarely brought up in the media. We most often see "Japan kills whales" and the facts do not go further than the number of the animals taken (moreover, that number is sometimes rounded to "thousands" without the details on how many hunting seasons or what year of history this information reflects).

This generalization is both the reason for and the result of the concept known as the "super-whale" among those who do research in the fields related to the whaling impasse. The term was first used by Arne Kalland in his article "Management by totemization: whale symbolism and the anti-whaling campaign" (1993) referring to the phenomena of characteristics belonging to different species of whales being integrated into one image of the "super-whale". Kalland explains that the "super-whale" – that is the biggest mammal (which is only true about the blue whale), has the largest brain of all animals (the largest brain belongs to the sperm whale), makes pleasant sounds (this is a reference to the humpback whale "song"), is curious about humans and "friendly" (somewhat true about the gray whale) and is endangered – simply does not exist (1993). According to the researcher, this construct came to dominate popular imagination representing all species of cetaceans. The article was written at the beginning of the 1990s, but the "super-whale" did not disappear from the imaginary horizon. It is one of the most important elements of today's common Western perception of whaling that effectively impedes any progress of Japan or any other nation on the way to making the practice of whaling acceptable.

The concept of the "super-whale" or comparable ideas were brought to my attention by many of this study's participants, who mainly expressed frustration at the fact that the West seems to be preoccupied with protection of endangered species and because of that, they attack the Japanese whalers. But the species of interest in Japan do not fall into that category (personal communication 2016–2019).

In parallel with conducting interviews I was doing desk research, following whaling-related news, and talking to friends and acquaintances about Japanese whaling a great deal. In all of these instances of coming into contact with the discourse on whaling, the image of the "super-whale" would repeatedly emerge. Although casual conversations with bystanders provide only anecdotal evidence, it bears noting that most dialogues I had on whaling with people who were not themselves involved in the issue, went according to almost the same scenario. "What is your research on?" they would politely ask. My vague answer would be, "The politics and anthropology of Japanese whaling". "Oh, yeah, what they're doing is so terrible!" they would convincingly proceed. "What do you think is so terrible about it?" would my next question be, asked with genuine interest. "Well, whales are endangered!" "But not all whales are endangered and the Japanese hunt those that are not". "Oh…" And with that, more often than not, anti-whaling arguments would end. If not – other "super-whale" attributes could come up. Such as "the whale" being smart – "smart" in itself being a complicated concept, and there being only some ongoing research on it for certain species of cetaceans (the bottlenose dolphin has the highest among cetaceans brain-to-body mass ratio, but the theory that this is an adequate measure of intelligence in animals is debated). Or humans having a special bond with whales, which could be considered true, but in a post-scientific way – we do seem to have a special bond, but this bond is unrequited. It was constructed by humans in the past five decades or so and has nothing to do with cetacean biology.

Starting from answering that one *what* question and recognizing the construct of the "super-whale" in action, the more general idea of misconceptions dominating the whaling discourse rings increasingly true as one's investigation progresses. It is a fairly simple, but undeniable observation that some of the most avid whaling protesters, as well as articles that promote anti-whaling ideas, are not basing their arguments on solid knowledge and understanding of Japanese whaling and the people behind it.

This research was preoccupied with some other *what* questions, that in my opinion are not often given due attention in the context of the multifaceted issue of Japanese whaling. For instance, the seemingly straightforward and essential questions of *what* values are behind it, *what* they think about it, *what* they are doing about it, and *what* they want to do next. The ones that would put the pro-sustainable whalers in the center of a blank page, possibly eliminating some of the biases and stereotypes, and covering the gaps that exist in perception of whaling. Questions of this kind shift the focus from searching for ways to fight and fix, to being curious and kind toward perspectives we do not ourselves hold.

Getting ahead of the narration, once a lot of the *what* questions were answered in the course of this research, giving way to a more accurate understanding of the Japanese people who defend the pro-whaling stance, the *why* do they act the way they do question disappeared by itself. As the Japanese say – *naruhodo*, which can be roughly translated as "I see" when you confirm your understanding of an issue.

The social landscape of Japanese whaling

One of the most important *what* questions that is often left unanswered or rather unasked, is about the actors in the scene of Japanese whaling. *What* kind of institutions, organizations and people represent Japan in phrases like "Japan defies the world"[15] when it comes to whaling? Who are the people working behind the scenes? And *whose* position is protected and prioritized by the government of Japan, sometimes at the expense of the country's international reputation?

As mentioned throughout this book, generalizations and simplifications are routinely used by the English-language media when it comes to Japanese whaling, and we see this tendency manifest itself in the treatment of Japanese whaling-related actors as well. They are mentioned in passing as if it is assumed that the reader is aware of who is who in Japan when it comes to commissioning whaling ships, conducting research or small-type coastal whaling (STCW), and other related procedures and activities. Perhaps, given the specifics of the genre, a certain lack of details in the media can be understood if their focus is not on actors, but on their actions. However, academic articles and semi-academic publications[16] on whaling also do not shed enough light on these blind spots. In academic literature on Japanese whaling the focus is often somewhat narrow. Again, due to the specifics of the genre and the complexity of the problem, most existing academic articles and books look into a particular aspect of Japanese whaling from a certain angle. It could be the culture surrounding whaling;[17] a specific type of whaling – STCW and large-type coastal whaling (LSTW),[18] pelagic or research whaling;[19] the international dimension of the issue, mostly embodied in analyses of the IWC and its work, the effect of it on Japanese policies and Japan's relations with and within this organization;[20] the animal rights perspective and the status of whales in the context of the issue;[21] or public opinion on the problem.[22] Researchers also devote special attention to particular developments of the issue. For instance, the Whaling in the Antarctic (Australia v. Japan: New Zealand intervening) case decided by the ICJ provoked a large number of academic publications,[23] and, of course, Japan's withdrawal from the IWC also did.[24] Although having a narrow focus is justified in most cases mentioned above, this creates a blurred background effect – seeing the big picture of the social landscape of Japanese whaling is not easy, as is understanding the relations between its different elements. Moreover, after having conducted this research, it became clear that certain important players on the Japanese whaling scene are almost entirely ignored by the existing texts pertaining to the problem. This section is included in this book in an attempt to bring more details of the existing diverse Japanese whaling scene to light.

To do that it was deemed best to turn to the emic perspective and let the people involved in it speak for themselves, explaining the entities and relations they perceive to be important in the context of Japanese whaling. The ethnographic tools of observation and casual communication during various whaling-related events were also used to complement the opinions and views of the study participants. Paying attention to the processes happening in interactions between the various actors in Japanese whaling and exploring their shared understandings revealed new nuances and added intricacy and depth to the overall social picture of Japanese whaling.

Starting from the organization that is, perhaps, most often mentioned in the context of Japanese whaling, the Fisheries Agency of Japan (*Suisancho* 水産庁) is an important actor on the Japanese whaling scene on the national government level. Interestingly, many of the texts that talk about it regardless of their pro-, anti-, or neutral orientation toward whaling frequently use the word combination "the powerful Fisheries Agency".[25] For an unprepared reader the word "powerful" or similar in meaning context associated with this institution followed by an absence of further explanations might result in confusion and belief that the Fisheries Agency somehow has more power than any other governmental entity. However, no part of my research provided any evidence of this. The Fisheries Agency of Japan is a part of the Ministry of Agriculture, Forestry and Fisheries. It is staffed similarly to other Japanese government units and operates in accordance with the standard procedural requirements.

Another governmental entity that is involved in the issue of whaling is the Fishery Division (*Gyogyoshitsu* 漁業室), which English-language materials rarely mention. On the international level of the whaling problem, the Fishery Division is very involved, as it is a part of a department at another ministry – the Economic Affairs Bureau of the Ministry of Foreign Affairs. It is in consultation with the Fishery Division that the reactions to the international whaling-related developments are being negotiated within the government of Japan.

As per the names of both these entities, they do not deal exclusively with whales and whaling, but with marine resources in general. This supports the argument that in Tokyo this issue is not seen as an exclusively international problem, or as an environmental concern. For Japan, this is first and foremost a resource management issue, and that is reflected both in the structures responsible for it, the composition of the Japanese delegation to the IWC throughout the years when Japan was a member of the organization,[26] the country's official statements, as well as what the people involved were saying about it in the course of this research project.

Since the names of the agencies that are involved in Japanese whaling do not presuppose a focus only on that one issue, the question arises about whether there are specific people whose work is entirely dedicated to this kind of marine resources. Within the Fisheries Agency at MAFF, there is a whaling unit called Whaling Affairs Office abbreviated to WAO (*Hogei no bushitsu* 捕鯨の部室). The number of people working here fluctuates, but usually, there are eight to ten staff members responsible for this office's work. All of them were interviewed during this project, individually or in small groups, and interacted with at several whaling-related activities and events.[27] Excluding the two higher managerial positions, other members of the team are young public officers in their 20s and 30s.

The Fishery Division of MOFA is much smaller than the Fisheries Agency of MAFF and does not have subdivisions for different categories of marine resources. The whole unit has six-seven officials working there, and at any given moment only one or two people are focusing specifically on whaling. One government official, who was formerly working at the Fisheries Division and was mainly responsible for whaling-related work was also interviewed for this research.

All civil servants in the Japanese government rotate between posts within their ministry every few years. The exact time depends on the circumstances, and at times returning to the same office is also possible. This process is the same both at the WAO at MAFF and the Fisheries Division at MOFA. The rotation is intended to provide the opportunity for officials to learn and gain experience in various fields, as well as to exclude the possibility of misconduct (*fusei* 不正) and abuse of one's position. One personal account of such a change of posts is as follows:

The first IWC meeting I participated in was the one held in 2007. After that I did whaling related work for 3 years, then for 2 years I was involved in fisheries cooperation, after that I was doing completely unrelated work, and then a year ago I came back to whaling again.

(a former WAO officer, personal communication, July 2017)

Another WAO officer shared that by the time he was interviewed for this study, he had served at WAO for 13 months as an assistant director, prior to that worked at the Fisheries Management Division of MAFF specializing in tuna management, as well as with the Economic Partnership Agreements team within the International Affairs Department of MAFF (personal communication, July 2017). Other members of WAO shared similar experiences. They were all transferred to the office without much prior knowledge of whales or whaling specifically, although all of them were educated in marine resources management or related fields. It is possible to express one's preference on what office they would like to join, however, this does not guarantee placement, and none of the people interviewed had indicated WAO as their first choice before they were assigned their whaling-related posts.

The same was confirmed by a participant who was a former officer at the Fishery Department for almost four years, but when I interviewed them was an assistant director with the Arms Control and Disarmament Division, Disarmament, Non-Proliferation and Science Department at the Ministry of Foreign Affairs. In their words *"It is very normal for MOFA to have people move from one issue to another without any relevance to their previous background"* (personal communication, February 2018). Their experience was as follows:

I was in the Fishery Division of the Ministry of Foreign Affairs for a long time – 3 years and 11 months. Usually at MAFF people change posts every 2–3 years. But when I was at the Fisheries Office there was the ICJ case[28], so everyone who was there at that time had to stay until that was over. Including myself.

A similar professional path was shared by a Senior Fisheries Negotiator Hideki Moronuki, who was for many years interchangeably serving at the Fisheries Agency and as a representative of Japan at the Food and Agriculture Organization of the United Nations (FAO) (personal communication, January 2017). In the course of this research, he was coordinating the work of the WAO and was Japan's Alternate Commissioner to the IWC in 2018. In 2023, he no longer serves at WAO.

This information is important in light of the image mentioned above of "the powerful Fisheries Agency" and the perception that is sometimes encountered in whaling debates that the people involved in the issue on the governmental level in Japan are so ardent about preserving whaling because of their own career interests (Blok, 2008). This was found difficult to prove in the course of this research. The Japanese policy of frequent rotation of staff in public service is the same across different ministries, MAFF and the Fisheries Agency are no different in this regard. The rotations make it highly improbable that anyone builds a government career based on whaling specialization alone since one necessarily gets appointed to other offices. Should Japan stop whaling altogether the government employees involved in the issue at that moment would just be transferred to other offices and departments, and most of them already have experience working with other marine resources or even in completely different areas. This is true for both higher-ranking officials and those who are only at the beginning of their careers, and for whaling as for any other area of governance.

The volatility of the issue of whaling in the international arena makes the work of both the WAO at MAFF and the Fisheries Division of MOFA challenging. As one staff member at WAO shared, *"Not all divisions (of MAFF) are facing some kind of criticisms, the reactions of the foreign countries differ depending on the resource, so that does have an influence on our work"* (personal communication, July 2017).

Another effect of the Japanese whaling policies having international resonance is that the WAO at MAFF works in consultation with the Fisheries Division of MOFA. This is also the case with other marine resources, but because the international community tends to react so intensely to the changes in Japanese whaling, the whaling-related officers at MAFF and MOFA have to coordinate their activities very closely, this was especially the case before Japan withdrew from the ICRW/IWC. As the nature of these ministries' work and their goals are inherently different, they view the issue from different angles – per their mandates, MAFF is predominantly preoccupied with the national dimension of resource management (although this ministry also has an International Affairs Department), and MOFA's priority is Japan's relations with other countries. The need to reconcile these interests in the unstable context of the whaling controversy creates extra pressure for

Figure 5.1 A policy explanation event held by the Whaling Affairs Office with attendees from the Japanese whaling community, including journalists regularly covering whaling, businessmen, local governments' officials, university professors, etc. MAFF Headquarters, Chiyoda, Tokyo, January 2018. Photograph by the author.

Figure 5.2 Food served at the policy explanation event held by the Whaling Affairs Office. Includes deep-fried whale meat (pictured in the center). MAFF Headquarters, Chiyoda, Tokyo, January 2018. Photograph by the author.

the civil servants involved. This factor was likely making whaling-related communication and decision-making within the government more challenging, which was confirmed by the interviewees' testimonies. In the experience of a participant who formerly served at the Fisheries Division of MOFA, *"It was at times hard in terms of communication. I had to ask for details from the people at MAFF and then communicate it to our legal team, and there were cases when it was interpreted in a way different from what was expected."* (personal communication, July 2018)

In towns that have historically been involved in the whaling industry and/or have traditions of small-type coastal whaling and consuming whale products for food, there are also civil servants and elected representatives in local governments whose work is related to whaling. Naturally, the Japanese politicians in the traditionally whaling regions tend to have strong opinions on the matter, and pressure from them has influenced Tokyo's moves on that scene. The most important of these whaling towns were discussed from a historical perspective in Chapter 1, and their current state and role in Japanese whaling will be analyzed further in this chapter.

Another important actor in the Japanese whaling scene is the Institute of Cetacean Research (ICR). It is a non-profit organization founded by the Ministry of Agriculture, Forestry and Fisheries of Japan in 1987 – the year after the moratorium on commercial whaling came into force. As specified in its name, the ICR specializes in research of cetaceans, but the organization's mandate also mentions "other marine mammals, as well as studies on related international matters".[29] Additionally, it coordinates the work of other institutions and individuals who specialize on research related to cetaceans.[30] Although it was working for a long time before that, it was officially nominated as a designated institution for the research of whales by the Fisheries Agency in 2017 and that mandate was extended in 2019 after Japan withdrew from the International Convention on the Regulation of Whaling, making the ICR the current primary institution promoting sustainable whaling in Japan's territorial waters and EEZ through research. The ICR became notorious in the international media, because of its role in the Japanese research whaling programs in the Antarctica and North-Pacific, which both included lethal sampling methods. As mentioned earlier, the programs were discontinued after Japan withdrew from the ICRW and the IWC.[31] The last research whaling trip to the Antarctic Ocean was concluded on March 31, 2019, with the vessels' return to the Japanese port of Shimonoseki. Already before that, at a briefing session at MAFF in January 2019, as well as at a whaling-related policy explanatory session at the Liberal-Democratic Party headquarters in February 2019, it was confirmed that research on whales and whaling will be sustained, albeit based on a different design (personal observations). The new research programs are now underway, but they attract less international criticism as they are now conducted exclusively on Japan's territory. With this, the role and position of the ICR on the Japanese whaling scene is not likely to change significantly in the near future.

Another organization that plays a significant role in whaling is the Japan Whaling Association (JWA). It was mentioned and described in the second section of Chapter 1 when I borrowed the JWA's chronology of whaling in Japan to partially base the discussion of the historical background of the issue on it. The JWA was founded in 1959 as a non-profit foundation, and then reformed in 1988 as a private organization with the main goal of fighting for the resumption of commercial whaling.[32] It was indeed actively working to achieve this up until Japan withdrew from the IWC, following which its objectives turned to supporting the whaling industry.

The JWA's activities currently include consolidation and dissemination of whales and whaling-related information, as well as organization of educational and promotional projects themed around whaling and whale-based cuisine. Some of these are planned and executed by the JWA alone, but most involve representatives from the ICR, as well as other institutions, smaller NGOs, and individuals (some of the most prominent ones will be mentioned further). Many of the events led by the JWA and other organizations were visited and observed in the course of this research. These included whale-meat cooking classes, promotional events at the Yokohama wholesale fish market, educational classes at schools, whale-themed-festivals, whale-meat themed food festivals, etc. The meaning and purpose of some of these will be discussed further.

It is possible that the methods and scale of the JWA's work will change somewhat as Japan adjusts to its new status of a country that whales for commercial purposes. However, it is likely that the essence of the organization's work will stay the same. In the opinion of the JWA's representatives and some other pro-sustainable whaling activists, the knowledge on whales and whaling in Japan is "under-satisfactory" and the consumers' interest in whale products leaves much to be desired (personal communication, 2016–2023). Resumption of commercial whaling means a change

Figure 5.3 A school lunch served in an elementary school at the end of an event organized jointly by the representatives of the Japan Whaling Association and Women's Forum for Fish, where information on whales and whaling was shared with the students and their parents. The menu includes whale meat soup and deep-fried whale meat. Toshima, Tokyo, January 2019. Photograph by the author.

in the means of delivering whale meat to the market, and, most likely, the available quantities of it. It does not, however, mean that it will automatically become more popular than it used to be, given among other reasons, its price that is still relatively high and the specifics of its preparation for food. Changing this situation is likely to become the focus of the JWA's future endeavors.

Kyodo Senpaku is another name that is often encountered in whaling-related oral and written debates. Kyodo Senpaku Kaisha Ltd. is a private company that has been known under this name since 1987, but it is linked to its mother company – Nihon Kyodo Hogei – and as such it has existed since 1976. Nihon Kyodo Hogei was formed as a result of a merger of several whaling companies that were active in the industry prior to the 1970s but were forced to join forces in the wake of failing whale stocks and consequently shrinking businesses.[33] For more than 30 years from its foundation in 1987 Kyodo Senpaku was leasing its whaling vessels to the Ministry of Agriculture, Forestry and Fisheries and the ICR for the conduct of the governmental research whaling programs and MAFF and ICR were the only entities the company worked with this whole time. Although the media kept calling Kyodo Senpaku a whaling company, strictly speaking, it only started being one again in 2019 when Japan restarted commercial whaling.

As briefly mentioned in Chapter 1, the ICR, Kyodo Senpaku, and the JWA all have their head offices on the same floor in the same building in one of the quieter areas of the Toyomi district in Tokyo. Kyodo Hanbai – a daughter company of Kyodo Senpaku that used to sell the by-products of the research whaling programs conducted by the ICR and is now tasked with selling meat from commercial whaling – is the only division of the three institutions that is located in a busier area of Japan's capital for logistical reasons. Such physical closeness provides for close coordination of activities between ICR, the JWA, and Kyodo Senpaku, and serves as additional proof of how interconnected their work is.

The above closes the list of large-scale institutions in Japan, whose work is primarily related to whales and whaling. This, however, does not cover the whole of the social landscape of Japanese whaling. Its other elements are smaller in size, but at times their impact on the issue is comparable to the big organizations.

One such element is the Japan Small-Type Coastal Whaling Association (JSTWA). Its members are mostly representatives of the towns already mentioned several times in this book – Taiji, Wadaura, Ayukawa, and Abashiri – and most of them are fishermen and whalers themselves. As per the organization's name, its focus is on STCW and issues specific to this type of activity. Japan was continuously applying for the IWC to recognize STCW as a particular whaling activity with long traditions and significance for community-building and the identity of the whaling towns. However, the IWC never accepted Japan's proposals, and no catch quotas for Japanese STCW were allocated on the IWC level. The JSTWA was participating in applications for the special STCW quotas working along with the government. JSTWA is overall an active participant in the Japanese whaling scene, with its representatives regularly visiting Tokyo for whaling-related meetings and events (personal observations, 2016–2019).

Another organization undoubtedly worth acknowledging here is Women's Forum for Fish (WFF). It is rarely mentioned in current whaling-related literature, yet its impact on the Japanese whaling scene is quite tangible, particularly in grassroots mobilization for pro-sustainable whaling campaigning, as well as whaling-related education. The organization is small, having only its head and main representative Yuriko Shiraishi, and director and chief editor Akiko Sato as permanent members, and a number of people who intermittently lend their professional skills, for example, dieticians, nutritionists, government officials, etc., to support WFF's projects. Shiraishi has been involved in pro-sustainable whaling campaigning since 1993. When she learned of the existence of this problem in the early 1990s, the activist was surprised to find that as Japan was defending its position in the international arena, *"the voice of the* [Japanese] *public was nowhere to be heard"*.

> The idea that it should not look like there are only a few people in Japan who are interested in this [issue] was the starting point for me.
>
> (personal communication, April 2018)

Shiraishi recognizes that with time the domestic public's interest in the issue was dwindling. However, to this day she believes that the support of Japanese whaling has to come from the people and still sees her mission in mobilizing popular support for the cause, since without it Japanese whaling is *"doomed to be forgotten" (Ibid.)*. In her words,

> whaling problem is a people's problem (kokumin no mondai 国民の問題) and we cannot let go of it.
>
> (Ibid.)

"Letting go" of whaling would be, in Shiraishi's opinion, tragic for the country, as it struggles with its self-sufficiency rate and food security. She also believes in the importance of whaling traditions for cultural reasons.

WWF first started by publishing materials on the nutritious value of whale meat and ways to cook it in local newspapers. But soon the women recognized that it would be more efficient if they could provide a more hands-on experience – that is how their focus shifted to live presentations, whale products-based cooking demonstrations or classes, interactive classes in primary and middle schools where children are taught about whales and whaling, educational and cooking events for mothers (WWF believes that *"whaling issue is a mothers' issue – if they won't cook, people won't eat"* (personal communication, April 2018), whales-themed poetry contests, as well as large-scale whale festivals, where whale products are sold, whale-based food is offered to attendees, and whale-themed performances are presented to the public.

WWF cooperates with both JWA and the ICR in organizing their events and I had an opportunity to attend a number of them while in the field. Shiraishi and Sato have been working hard to expand their reach for the past 30 years, continuing

Figure 5.4 A nutritionist demonstrating how to cook whale meat to a group of mothers. An event organized by Women's Forum for Fish. Tokyo, October 2018. Photograph by the author.

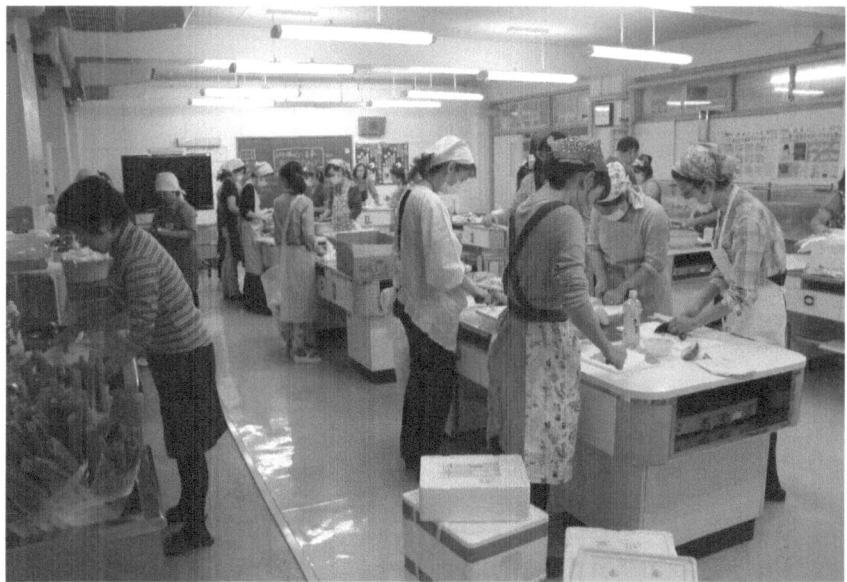

Figure 5.5 A group of women cooking whale soup as a part of a school event organized by Women's Forum for Fish. Tokyo, October 2018. Photograph by the author.

even through the hiatus of the COVID-19 pandemic when they shifted to offering their events virtually. Further demonstrating their commitment to the issue, the two women have also been regular attendees of the IWC Committee meetings, flying to different locations using their personal funds.

There are also individuals who demonstrate a passion for the issue of whaling and play an active role in the Japanese whaling arena. They usually have a long history of lived experiences related to whaling. Most of them operate businesses that heavily depend on whale products or have ambitions to build such businesses. I had an opportunity to meet some of these people and they were all knowledgeable of the history and current developments in the whaling controversy and had strong opinions on the issue.

One of them is Junko Kojima, who took over the ownership of "*Kujirakan*" restaurant in Shimonoseki city that was opened by her late father decades ago. Kojima runs the place with her mother and is also the main chef. She loves talking about the whaling issue with those of her customers who express curiosity, and informing her regulars is one of the reasons that make her feel the need to be aware of the latest developments in Japanese whaling. Like the women from WFF, Kojima also attended many of the IWC Committee meetings on a self-funded basis (personal communication, September 2018).

Another businessman whose restaurant's menu is almost exclusively based on whale meat is Hajime Ishikawa. The young entrepreneur opened "*Himitsu Kujira*" ("Secret Whale"), currently "*Kujira*" ("Whale"), after having majored in fisheries in college and working in distribution of whale products for several years. Through

Figure 5.6 Women's Forum for Fish presenting on Japanese whaling to a group of attendees. Tokyo, March 2018. Photograph by the author.

Figure 5.7 Attendees of an event jointly organized by Women's Forum for Fish and the Japan Whaling Association in the background. Deep fried whale meat on the foreground. Photograph by the author.

Figure 5.8 Deep fried whale meat cooked and served to the attendees as part of an event jointly organized by Women's Forum for Fish and the Japan Whaling Association. Tokyo, March 2018. Photograph by the author.

Figure 5.9 Women's Forum for Fish presenting on Japanese whaling and cooking whale meat to a group of school students as a part of an event jointly organized by Women's Forum for Fish and Japan Whaling Association. Tokyo, January 2019. Photograph by the author.

Figure 5.10 Schoolchildren standing in the shape of a Sei whale – an activity led by the representatives of Women's Forum for Fish in a joint event organized with the Japan Whaling Association. Tokyo, January 2019. Photograph by the author.

previous work, Ishikawa met the owner and CEO of Gaibo Hogei[34] – Yoshinori Shoji – and was inspired by Shoji's passion for preserving the culture of eating whale meat. Currently, all whale products for the restaurant are sourced from Gaibo Hogei. Ishikawa's vision for his restaurant and whale products-based cuisine, in general, is to make it "*oshyare*" or stylish, novel, unique, and on the expensive side. He sees his target customer base as young people coming in for dates and special events. Different from other restaurants where whale meat is served, "Kujira" looks and feels like a fashionable and modern spot, the kind where you are not just eating, but having a dining experience (personal observations, August 2018).

Ishikawa's approach to positioning whale cuisine as a worthy addition to Tokyo's sophisticated dining scene is somewhat new in the world of Japanese whaling. But there are signs that it might be picked up by bigger players in the Japanese whaling arena soon. Whale meat is traditionally seen as a product older generations of Japanese people are familiar with from the post-war years and are supposedly nostalgic about, while a younger crowd would have less interest because of the limited availability of whale products during their lifetime and, hence, barely having exposure to whale-based dishes. However, the future of whaling depends on

Figure 5.11 The entrance to the "Himitsu Kujira" restaurant. Tokyo, February 2017. Photograph by the author.

Figure 5.12 Whale sashimi served at "Himitsu Kujira" restaurant. Tokyo, February 2017. Photograph by the author.

the interest toward whale meat from that latter group, and targeting them seems a viable strategy for the survival of both the whaling industry and the culture this industry feeds.

Another person often present at both formal and informal whaling-related events is a former intern at JWA, later an employee of Kyodo Senpaku, and currently an importer of Norwegian whale meat to Japanese markets and an active member of the Japanese whaling community, Hirohiko Shimizu. Having high ambitions for the development of the whaling business in Japan, Shimizu also organizes some unofficial gatherings himself. He has boarded Norwegian whaling ships to help the crew and do research for his business many times, and in 2021 he spent several weeks doing the same onboard a Japanese STCW boat (personal communication, December 2021).

One more person, whose work demonstrates the dedication and versatility among the members of the Japanese whaling community is the work of Kiyokazu Yoshimura. Now a communications specialist with the JWA, in 2021 he spent a whole whaling season on board a commercial whaling ship helping with flensing and other parts of operations. Yoshimura has a lot of practical experience with whaling, as he was a crew member prior to moving to whaling-related office work.

The ease with which some representatives of Japanese whaling exchange their business suits for lifejackets and roll up their sleeves to work with whale meat is also emblematic of the close ties within this community. People know each other's stories and of each other's experiences and work together accordingly.

Figure 5.13 Hirohiko Shimuzu, searching for a whale in Norwegian waters, 2019. From personal archives of H. Shimuzu.

Figure 5.14 Kiyokazu Yoshimura on board a commercial whaling ship with a body of a Sei whale, the Pacific Ocean east of Hokkaido (within the EEZ of course), October 2021. From personal archives of K. Yoshimura.

In the context of the acrimonious anti-whaling attitudes expressed toward Japan by non-whaling nations, an important question is whether or not there are any actors supporting the anti-whaling norm inside the country. The emic understanding is that people in Japan are, perhaps, not all pro-whaling, but mostly not anti-whaling either. Confirming that, a WAO officer shared the following:

I have on different occasions asked my friends about their thoughts on whaling. And there were several people, who said it's a shame (sheimu da シェイムだ). But most of them said that if there is plenty of research done and it shows that it's alright to take that many, it should be fine (betsu ni ii n ja nai no 別にいいんじゃないの). And more of them than I expected had this [latter] opinion. Maybe there are people in Japan who are against whaling, but they do not go as far as acting upon it. I would say that almost no people take anti-whaling action here.

(personal communication, July 2017)

The data collected in the course of this research agrees with this assessment. However according to the 2019 polls conducted by MAFF, close to a third of the Japanese public (27%) found the country's decisions to withdraw from the ICRW/IWC and to start commercial whaling in its own territorial waters and EEZ negatively (choosing "do not rather value them" or "do not value them"),[35] active campaigning against whaling within the country is rare.

Although the intention of this book was to give a balanced account of the social elements on the Japanese whaling arena, it proved challenging to identify and engage the representatives of the anti-whaling side within Japan. The only local organization in Japan that specializes in cetaceans and campaigns against hunting and capturing these animals for the entertainment industry is Iruka and Kujira Action Network (also Dolphin and Whale Action Network or IKAN). Although some refer to IKAN as a "domestic anti-whaling group" (Kolmaš, 2021; personal communication, 2019), it in fact has one permanent member who is also the executive director – Nanami Kurasawa, who I had a chance to speak with while working on this book. IKAN was established in 1996 with the mission of having cetaceans properly protected and managed and the main way Kurasawa and her supporters work toward this goal is through dissemination of information.[36] Since its founding IKAN has cooperated with other organizations, most of them local chapters of big international environmental NGOs (ENGOs), such as Greenpeace and International Fund for Animal Welfare, as well as with individuals, including scientists and journalists sympathetic to IKAN's cause. Demonstrating strong dedication to her work, Kurasawa often attended the IWC Committee meetings, including the one I attended in Brazil in 2018. Although Kurasawa herself is a long-time vegan, she underlined that her ambition lies in stopping discrimination against marine mammals in Japanese legislation and supporting the animals' just treatment rather than insisting on the ideas of vegetarianism or veganism (personal communication, June 2017). She also admits that she supports the idea of STCW in traditional whaling towns, as long as it is done sustainably (Kolmaš, 2021).

Another Japanese NGO that campaigns against the hunting of cetaceans among other issues related to animal welfare and environmental protection is Life Investigation Agency (LIA) established in 2010.[37] LIA sponsored a trip to Brazil for one of the anti-whaling protesters that I met at the IWC Committee meeting in September 2018, and that was how this organization's existence came to be mentioned here. LIA's anti-whaling activities are sporadic and mostly online-based, and it is a fair example of what anti-whaling within Japan looks like in general. It is mostly represented by organizations who include cetaceans in their broader agendas of wildlife protection, and the level of attention paid to the issue corresponds to that.

At the invitation of Nanami Kurasawa, I had an opportunity to attend one event where the main purpose was to share a critique of the Japanese whaling policies that were current at that time. It was organized in cooperation with the Network for Biodiversity Conservation Legislation and held at one of the National Diet buildings. Most of the presentations and the discussion centered around the research whaling programs, now discontinued, and marine wildlife-related policies. The announcement of the event put its capacity at 40 attendees, but it had only about 15 people present the day of, including five speakers and the media. This was the only domestic event that I knew of and attended during my fieldwork, where participants held a more critical stance on Japanese whaling.

The scale of this event demonstrated that the *non-pro-whaling* forces in Japan are less consolidated than pro-sustainable whaling ones and attract less support domestically. Notably, "non-pro-whaling" is a better word choice to denote this group than "anti-whaling" would be – the stance that not a single whale of whatever species can ever be taken under any circumstances is not a popular one in Japan.

Overall, the social landscape of Japanese whaling is rather stable. Excluding government officials, who necessarily leave their whaling-related posts for other departments, as prescribed by how the public service system operates in Japan, many of the individuals and groups at the heart of the Japanese whaling complex tend to have long experience with it and usually stay concerned about this issue for many years. Some additional elements – individuals or groups – join for some time and soon leave the scene, their impact on the whole system hardly noticeable. I identify myself as one such element – I was someone who entered the proverbial field one day and started regularly speaking to people and coming to whaling-related events, from restaurant gatherings to official government presentations. I was introduced to people, guided around, indulged with detailed explanations, and invitations. My presence was noted and people, no doubt, talked about this foreign woman, fluent in Japanese (which is still not very common in Japan), asking questions. For a short while, I was an active element of the social circles of Japanese whaling. Through this book, I will stay present for some time, albeit not as active and not in person, leaving my small mark in the social landscape of Japanese whaling. However, the people living and breathing whaling, day in and day out, tend to stay active in it for decades, and they are the ones who form the committed core of this community.

The communication between different actors and groups of actors on the Japanese whaling scene is regular and timely. This was mentioned by the interviewees and was observable in the field. Close coordination is certainly happening on the governmental level – a staff member of WAO at the Fisheries Agency assured that *"it goes without saying that as soon as something new happens, we come together with the people from ICR [the Institute of Cetacean Research], and if there is something in relation to other countries, we'd have a meeting with the MOFA people"*, and some of his colleagues seconded that. But there are also strong government-non-government connections, and outside of the government, the whaling-related people are all linked as well. Most people who took part in this research project knew each other, or at the very least knew of each other.

A former director at WAO shared that frequent direct communication with the people who are the main beneficiaries of his and his colleagues' work was for him the most encouraging side of serving with this office:

In case of whaling we have a lot of chances to deal with the fishermen directly. Other divisions of the Fisheries Agency, since it is a national institution, usually deal with local governments or some other smaller agencies. In our case there are only a few people involved in the issue on the ground, so in a lot of cases we communicate with them without any third parties. That is something that is not a part of the work of other offices and divisions. We go to places like Taiji and Hokkaido and talk to people who actually do STCW [small-type coastal whaling]. They usually have a chance to ask us directly about international negotiations and express their opinions. With new national policies it's the same – we always know what they think directly from them. It is interesting to see these people bring to life the policies and

guidelines we, as government officials, work on, and it is very satisfying to learn about it when it goes well and brings positive changes to their work.
(personal communication, July 2017)

This quote reflects how close the different actors on the Japanese whaling scene are. Not everything in this domain is organized in the best possible way, according to the emic perspective – certain people, for instance, expressed concern that they were not notified of some events, or that explanatory sessions were not organized soon enough after changes in whaling-related policies were introduced (personal communication, 2016–2019). However, from the etic perspective these perceived failures in communication provide evidence of how strong the connections are within the Japanese whaling-related community. If a participant was not notified of an event, they would still end up knowing about it. If someone is not satisfied with the timing of a policy explanatory session, it shows that such sessions are expected as a matter of course.

Junko Kojima ("*Kujirakan*" restaurant owner and chef, Shimonoseki) also shared that being a part of the unofficial nation-wide group of people, whose work is related to whaling gives her strength and motivation, because *"you attend one of the information sessions or other gatherings* [Kojima sometimes flies to Tokyo specifically to attend these], *and you know everybody, and it is a chance to see these people and talk to them, and it is nice"* (personal communication, July 2019). This sentiment was not voiced as explicitly by other interviewees, but judging by my fieldwork observations, Kojima is not the only one enjoying the community spirit in the context of the Japanese whaling scene.

Continuity in Japanese whaling

There are certain characteristics of Japanese whaling that in the course of this project were identified as key to understanding its workings. These underlying patterns and processes that shape the Japanese whaling as we witness it today, are not necessarily unheard of in the context of the whaling discourse, but they are given more detail and granularity here below.

One of such pillars that supports the Japanese whaling-related people in their quest to preserve whaling is *continuity*. This concept was already invoked earlier in this book in the introduction to Chapter 1 where the importance of history in the issue of whaling was touched upon. History's significance is undisputable in its role of providing context for any problem as it develops over time. In the case analyzed here, however, history is so integrated into the present Japanese whaling world that it is hard to tell at times where the threshold between history and the present is. Continuity is being established between what is of historical importance and what constitutes today's reality in Japanese whaling.

Continuity is a value that is being upheld and actively perpetuated by the pro-sustainable whaling side. This manifests through material ways of paying respects to history, namely carefully preserving any remnants of the past days of whaling, as well as constructing new monuments to it – in the broad and even metaphorical sense of the word. History and succession of traditions are a part of the Japanese domestic academic and media discourse on whaling – many articles talk about it directly or imply the significance of whaling history for how whaling should be perceived today and what future policies should reflect.[38]

The theme of history, traditions, and culture emerged in the interviews conducted for this project. The language used by the Japanese people involved in whaling grounds this practice in history and shows it as a part of Japanese culture – the words *hogei bunka* (culture of whaling) or *geishokubunka* (whale-based food culture) were used by every participant.

Throughout the long history of whaling in Japan there have naturally been changes in hunting locations, species targeted, as well as innovations in techniques and technology used for catching cetaceans. However, as the discussion in Chapter 1 shows, there is continuity and consistency in the development of whaling in this country. The locations where some of the chronological milestones occurred demonstrate that in certain areas of Japan people have kept to their whaling traditions for many generations.

I took several trips to two of the places that are mentioned in the JWA's chronology of whaling in Japan. There are more locations in Japan where whaling played a role in the past and in some cases is still important. But only three modern towns are mentioned in the JWA's list, and all three of them very early on – the roots of whaling in Japan are from there. The same towns are also often mentioned in the whaling-related media and academic sources in relation to the modern-day whaling controversy – demonstrating the idea of continuity within the issue.

The first two places mentioned in Chapter 1 – Taiji and Wadaura – are visibly devoted to their whaling history and present. In 2023, the streets of both towns are dotted with shops and restaurants serving whale meat-based dishes, and there are statues and other artifacts reminding visitors about these places' unique history and present specialization. Wadaura's train station and roadside station buildings currently host small whaling museums that feature items related to the town's involvement in whaling throughout the years.

Taiji would, perhaps, rank first in terms of how much of its looks and culture are still shaped around whales and whaling out of the whole of Japan. A lot of details speak to this here – from obvious representations like monuments to whales that are decades old, a harpooner statue built in 1998, and the Kyo Maru No. 1 – a 69-meter-long vessel retired from research whaling in the Southern Ocean – on display here since 2012, to small details like images of whales on bus stop signs, tiles, information boards and manhole covers.

There is a large Whale Museum in Taiji that provides its visitors with detailed information on the locals' relationship with whales (Figure 5.15). It opened in 1969 and its complex includes two main parts. The first is a three-floor museum building with expositions displaying whale skeletons and bones, whale earwax, baleen, as well as paraphernalia related to local developments in whaling. The second part is an aquarium where whale and dolphin shows are being held in a natural dammed area adjacent to the museum grounds. The establishment is popular with tourists from across Japan with some foreign visitors.

Figure 5.15 The statue of a harpooner in the forefront (built in 1998) and Kyo Maru No. 1 whaling ship in the background on the left (on display since 2012). Kujira Hama Park, Taiji, July 2017. Photograph by the author.

Figure 5.16 A manhole cover in Taiji. July 2017. Photograph by the author.

Figure 5.17 The sign at the "Taiji Whale Museum" bus stop depicts a whale-shaped bus and the main building of the museum in the background. Taiji, July 2017. Photograph by the author.

If we placed the arrangements/erection/installation/organization of all these whaling-related elements in Taiji on a timeline we would see that there are no long gaps between them. Japanese whaling originated in Taiji about 400 years ago, and regardless of the difficulties this town has had to face in relation with this practice on several occasions, from the early 17th century to the present there was not a period in Taiji's life when it was not associating itself with whaling.

Taiji has a number of whaling-related tangible and intangible assets registered as its cultural heritage with the town government, as well as prefectural and/or national governments. These include the Whale Dance that celebrates the bravery of whalers and is performed regularly at the town's and national events (designated as intangible folk-culture heritage by Wakayama Prefecture in 1970); the Grave of Whaling Founder – Chubei Yoritomo Wada, who, as mentioned in Chapter 1, started organized whaling in Japan (designated as a historic site by Wakayama Prefecture in 1959); the Memorial Tower for Whales erected in 1768 by a whaler named Hachibei Hama, who was regularly coming here to ask forgiveness for ending the lives of whales – creating a tradition followed by many generations of whalers that came after (designated as a historic site by Wakayama Prefecture in 1984); the Site of the Remains of the Rest House for Watchmen at Tomyozaki Point (designated as a historic site by the town of Taiji in 1982); the Smoke Signal Station – the spot where whale watchmen used to light a signal fire (designated as a historic site by the town of Taiji in 1983); the Ibuki Tree Couple – two big old Chinese juniper trees that are believed to be planted by a whale watchmen and are preserved by the local people as the symbol of Kandorizaki watching point (designated as a natural monument by the town of Taiji in 1983); the colorful distinctively painted Chaser Boats of the Old Whaling Era on display at the Taiji Whale Museum (designated as cultural property by the town of Taiji in 2014); Modern Whaling Guns – a collection owned and displayed by the Taiji Whale Museum (designated as cultural property by Wakayama Prefecture in 2007) (Taiji town, 2015a).

There is also the Whale Memorial at Kandorizaki Point constructed in 1979, which at the time of writing did not have official designations, however, it is included in the Taiji town government published "Taiji's Cultural Heritage" booklet (2015b), and it is a monumental piece embodying the fact that the people of Taiji highly value the role of whales in the life of their community. Every year on the 29th of April a Whale Memorial Day event is held here to express the locals' gratitude to the spirit of whales. In 2018 about 100 people, including the Retired Whalers Association members, the town's Fishermen Cooperative representatives, and town's officials, burnt incense in front of the whale statue to the accompaniment of sutras chanted by a Buddhist priest (Kumano Shimbun, 2018). In his address to the attendees of the event, a senior Taiji town government official Hironobu Ryuno stated that Taiji will continue to be engaged with whales – today and forever, all while cherishing the past. He also conveyed that he wanted to pass on Taiji's long history of close relationships with whales to future generations (Wakayama Hoso Nyusu, 2018). The mayor of the town, Kazutaka Sangen, also confirms that protection and

Figure 5.18 The Whale Dance. "Japan Heritage – Living with Whales – Tales of Whales and Men" event in Tokyo, February 2017. Photograph by the author.

development of Taiji as a dolphin and whale town is an official local government policy, and the town has plans to turn its whole territory into an international whale and dolphin research project in the future (personal communication, July 2017).

In addition to the Whale Memorial Day event, there are several more whaling-themed events held in Taiji every year. Taiji Isana Matsuri – "*isana*" is the old Japanese word for "whale" and "*matsuri*" stands for "festival" – takes place on the 14th of August. The festival was founded in 1987, about a year after the moratorium on commercial whaling came into effect. As stated on the official website of the town of Taiji, the goal of the event is to sustain and promote Taiji's history and culture of whaling amidst the international criticism of the practice (Taiji Town, 2015a). The central show of Isana Festival recreates a part of the process of net-method whaling in one of the town's harbors.[39] Other regular elements of Japanese summer festivals, such as dancing, eating snacks from a variety of stands and fireworks are also there to enjoy during the day. Whale-meat-based food is, of course, also offered to visitors among other options.

Another festival – Taiji Kujira Matsuri – "*kujira*" being the currently commonly used word for "whale" – annually attracts visitors to Taiji on the first Sunday of November. The festival features a folk dance parade that includes the above-mentioned Whale Dance, as well as other performances and a food bazaar with whale-product dishes and whale products for sale (Taiji Town, 2015a).

Most of the above events, as well as others that are not as heavily whaling-themed, such as, for example, a local marathon, have a person dressed as the Taiji

Figure 5.19 A Taiji town government staff member next to the Whale Memorial at Kandori-zaki Point. Taiji, July 2017. Photograph by the author.

town mascot to entertain the public. Some of Taiji's information boards, as well as many of the brochures for events organized by Taiji town, locally published posters, etc. feature the drawing of the same talisman creature. It is not uncommon for Japanese municipalities, as well as government agencies, companies, groups, clubs, etc. to have their own mascots for branding purposes, and Taiji's was officially registered in 2009. It is a character named Gonta that predictably resembles a whale.

The official website of Taiji is also a testimony to the fact that whaling is a part of the town's identity. Its header defines Taiji as "an eco-museum of whales and ocean". A drawing of a whale is also a part of the header. There are pictures of the town's whaling-related sites and drawings depicting whaling boats, whales, and dolphins all throughout the website.[40]

All of the above reflects intentionally sustained and cultivated continuity between history and the present-day reality in Taiji. Many towns have places of historical importance and items that are preserved for their importance to the locals', some have mascots and most promote their special cultural or geographical features. In the case of Taiji all of the details presented here above form a single coherent narrative built entirely around whales and whaling. They reveal that its townsmen take pride in the past of their birthplace. An impressive variety of means have been explored by the local community and government to highlight the importance of the relationship the town has had with whales for more than four centuries. The conversations I had with people who live and work in Taiji revealed that they also hope for the future where they will enjoy an equally strong connection with these animals, including through whaling and using whale products for the benefit of the community.

Another town mentioned in Chapter 1 is Ayukawa. Ayukawa displays similar tendencies of both material and immaterial expression of the town's long-standing ties with whaling. However, due to circumstances that are outside of human control, the visual evidence of the locals' attachment to whaling culture is now harder to come across here than in other "*kujira no machi*" ("whale towns"). Due to its location, Ayukawa was one of the places that suffered heavy damage from the Great East Japan Earthquake of 2011. An over 8-meter-high tsunami destroyed more than half of the town's structures, including 500 homes and the buildings of the local branch of Gaibo Hogei whaling company (personal communication; Fackler, 2011). The whaling boats that belonged to the company were washed up the coast and grounded – returning them to operation was a costly undertaking. The overall loss of material property in Ayukawa was dramatic, not to speak of the loss of 23 lives (Tani, 2012). The town's population plummeted as well, as younger inhabitants now had more incentive to move inland in search of opportunities. In these circumstances, many observers predicted that this natural disaster might become the end of whaling in Ayukawa, and even the locals themselves were skeptical about its recovery (Fackler, 2011; The Japan Times, 2011). At the same time the people here were still hoping to see the industry rebuilt, feeling that if it disappears completely, Ayukawa itself will disappear with it (Carney, 2014).

The fact that after the tragedy of March 11, 2011, the locals were still able to rebuild the town and get back to whaling speaks volumes about the meaning it holds to the people here. Twelve years after the tsunami, Ayukawa is back on track with active participation in the country's efforts to preserve, revive, and strengthen its whaling culture.

Representatives of all three towns discussed here – Wadaura, Taiji, and Ayukawa – are regular attendees and/or participants at various whaling-related events in Tokyo, including a concert and panel talk event "Japan Heritage – Living with Whales – Tales of Whales and Men" (日本遺産〜鯨とともに生きる〜くじらと人の物語〜) held on February 3, 2017; The 29th Gathering on Protection of Whaling Traditions and Culture (第29回「捕鯨の伝統と食文化を守る会) held on May 11, 2017; "6th meeting with the IWC Commissioner Morishita" (第6回森下 IWC コミッショナーを囲む懇談会) held on January 30, 2018, in the offices of the Ministry of Agriculture, Forestry and Fisheries; the "Joint meeting of the whaling committee and the parliamentary group on whaling" (捕鯨対策特別委員会・捕鯨議連盟合同会議) that was largely themed around the next steps Japan will take after withdrawing from the IWC held on February 2, 2019, in the Liberal-Democratic Party of Japan headquarters (observations, 2017–2019) and many others held during the time I was working on this research and beyond. Participation from the whaling communities is encouraged by the country's whaling-related institutions, as preservation of their whaling history and development of their whaling-related future is considered a priority. Regardless of the strong international opposition since the 1970s, Taiji, Wadaura, and Ayukawa are some of the Japanese locations where due to the collective effort of all interested parties, the abandonment of whaling history and culture was prevented.

The so-called "culture argument", which is closely related to the concept of continuity discussed above, has been one of the contentious points in the debate on Japanese whaling. Some Western texts acknowledge the fact that Japan sees whaling as a part of its unique food culture and culture in general. But most of those are pointing out the fact that even if this can be accepted as true, Japan's hunts in the Antarctic – its now discontinued research programs – started recently and there is nothing cultural to them.[41] Japan did struggle to come up with a clear explanation on how to incorporate pelagic whaling in the Antarctic into their understanding of the culture argument. Mostly, this reproach was avoided by the representatives or defenders of Japan's position.

Although Japan currently does not conduct research whaling in the Antarctic, it now engages in commercial pelagic whaling in its territorial waters and EEZ. Pelagic whaling is indeed relatively new and the whole point of it is using bigger, more modern ships to scale up production. But can one deny the overall importance of culture, and hence traditions and continuity, based on technological advances in a practice? The result of whaling, be it in a country's territorial waters, EEZ, the Southern Ocean or even imported from Norway or Iceland, is still whale products. Keeping with its centuries-long traditions, Japan will use all of it for food, without wasting much of the whole whale carcass. And using whale products for food is at the heart of that unique Japanese whaling culture that, according to the emic perspective, is worth being researched, sustained, and preserved. As long as sustainability is well-researched for the species and stocks targeted, no valid argument against it can be made (without going into the territory of not allowing animal products' consumption in general, which is outside of the scope of this research).

It needs to be noted here that Japan prioritized support and preservation of coastal whaling in its negotiations within the IWC during the time it was a member state. Japan applied for the IWC to approve certain quotas for its STCW more than 20 times – some applications were discussed but later withdrawn, most applications were denied. Considering the fact that more than 20 times in the framework of IWC Commission meetings means more than 20 years and many hours of work in order to edit those proposals to address the criticisms they were getting each time, it can be concluded that the Japanese government does not lack motivation and dedication when it comes to advocating for its coastal whaling on the international arena.

For those who disagreed with Japan's whaling in the Antarctic, positive news came on December 26, 2018. As briefly mentioned earlier in this book, announcing its withdrawal from the International Convention on the Regulation on Whaling and the IWC, Japan at the same time committed itself to discontinuing its research whaling in the Antarctic – as in accordance with the ICRW, whaling in international waters can only be conducted by countries that are members of an international organization that works for the regulation of whaling. This was a major shift in Japanese whaling policies, but it remains to be seen if with that the conversation on whaling will change toward more acceptance of Japan's stance. For now, according to a communications specialist from the ICR in Tokyo, the media requests and critical commentary coming their way decreased significantly (personal communication, March 2023).

Continuing the discussion I started in Chapter 1, let's look at what role history plays in the Japanese whaling debate. Looking at a present phenomenon using a historical approach always adds to the depth of understanding a problem. Going back in time and tracing how both the challenges and solutions whaling brought to the Japanese people continue to influence the present is a useful exercise. However, a long history is not only a reason why a community could and should continue a certain cultural practice. Nor it is only an excuse, as the anti-whalers see it, suggesting that not all traditions are worthy of sustenance.[42] Continuity offers a simple, but strong explanation for the feelings Japanese people hold for whaling. As was established in this book, depending on the region, people were connected to this practice, and through this practice to each other, for over four centuries, having its presence in their lives for generations. This prolonged exposure produced the *effect of normalization*. Seeing people buying, selling, and cooking whale meat, or eating it for lunch at school, or having it for dinner with your family while growing up paves the way for familiarity and trust.

Several of this study's participants shared stories of childhood experiences related to eating whale dishes. One of them is Hajime Ishikawa, owner of a specialized whale meat-based cuisine restaurant and a chef in his late 30s, who was featured earlier in this book, had his father introduce him to the taste of whale bacon:

> I used to eat whale bacon at home, I think I first tried it when I was around 5–6 years old, my dad was eating it as a snack with his drinks, and he let me try. It was delicious and that special taste (衝撃的な美味しさ) made a big

impression on me (impakuto no aru インパクトのある). And I know that for my dad that taste was very special (totemo tokubetsu とても特別) too.

(personal communication, May 2017)

Notably, Ishikawa is a native of Ibaraki Prefecture, which is not considered to be one of the traditionally whale-eating prefectures. This quote points at two aspects of the problem that are often denied or understated. First, whale meat is eaten and known all over Japan to varying degrees. It is true that certain towns and prefectures have longer histories with whaling and using its products. Whale eating culture is more familiar to people in these, which was confirmed by the representatives of the WAO at MAFF. But that does not mean that outside of these areas people do not eat or have not heard of eating whale meat, or that whale meat eating is as marginalized as it is at times imagined and presented in the anti-whaling discourse. Even in areas of Japan where whale meat is not a regular part of people's diets, people know of its existence on par with other regional specialties.

Another misconception countered by Ishikawa's early life experience, along with him now leading a successful whale meat-based restaurant business in the heart of Tokyo for the past eight years[43] is that only older Japanese people are interested in this problem, as they were "reluctantly" eating whale meat in the absence of other sources of animal protein after World War II.[44] The continuity in Japanese whaling traditions was not broken in the 1960s or 1970s and not even after the moratorium on commercial whaling took effect in the 1980s. Some Japanese people continue enjoying whale meat, and some of these people are middle-aged or young. As reported by Ishikawa, who maintains a log of all reservations and walk-ins to his restaurant, more than a half of his guests are not older people (personal communication, January 2019). This is just one story from one person, his words supported by his own business development efforts – that is the reason he keeps the log – so representativeness is not claimed here. However, as suggested by the phenomenological and ethnographic framework adopted for this study, individual impressions are valid, and for the goal of improving our understanding of whaling experiences in Japan they provide valuable context we would otherwise lack.

Before the Western anti-whaling campaigns hit Japan, eating whale meat was not seen as something that needed justification. It existed as a commonplace part of life. Keeping that in mind, it is easy to imagine that a belief that there is something wrong with it would meet more resistance in Japan than in Europe, where there was no whale meat-eating culture to speak of. If one considers the validity of the anti-whaling claim that a long multigenerational history of a practice is not reason enough to continue it, they should also consider that people who make that claim do not come from the tradition and background that they make this judgment about. The products of whaling have never been considered food in the West, let alone a food people have fond memories of, nor has it ever been the only meat people knew. On the other hand, the historic emic perspective shows us that the situation looked very different for the Japanese people throughout the four centuries of whaling, as well as in the 1970s when "save the whale" campaigns gained force, in 1986 when the moratorium on commercial whaling introduced through the IWC came into effect and still looks different now. In the 1970s, whale meat dishes were at the peak of their popularity in Japan, after the country had limited access to other kinds of animal protein for years after World War II. In the beginning of the 1980s, people in places like Shimonoseki were stocking up on whale meat in fear of diminishing supplies associated with the looming prospect of the moratorium on commercial whaling. In 2023 people in Japan still enjoy whale meat-based dishes in restaurants across the country and it is seen as normal.

The tangled roots of tradition and modernity in Japanese whaling are brought up here not to provide justification for saving this practice for future generations in areas where whaling existed for a long time. In line with anti-whalers and the official Japanese government position, a stance is taken here that having a history of whaling is both not enough and not necessary for a community to be interested in sustaining or developing this practice, provided that sustainability is carefully considered. However, as established above, from the emic perspective whaling is perceived as natural, nostalgic, and familiar. Continuity in whaling and consuming whale meat for food has a lot to do with these feelings.

Han-han hogei or reactivity in the Japanese whaling dilemma

One of the points made in the previous section brings us to another important pillar of Japanese whaling - reactivity. In psychology, this term is used to talk about the tendency to change one's behavior in response to being watched. It is observable when people's movements/actions/activities are being measured in laboratories (it is a factor that has to be controlled for when conducting research), or in daily situations when people behave differently in a public setting than they would if they were alone (French & Sutton, 2011). Reactivity is considered normal human behavior, and although it is studied by professional psychologists, it is also easily observable by non-professionals.

This section argues that reactivity as a concept can help explain many aspects of the Japanese actors' behavior within the issue of whaling. In International Relations theories the idea of international norms influencing countries' behavior is widely discussed. This is applicable to the case of Japanese whaling as well. However, the term reactivity was borrowed from psychology and brought into the discussion here, as the framework used in this research project represents an approach where individual and small-group levels are prioritized.

Judging by the conversations I had during this study, the representatives of the Japanese whaling scene are not all cognizant of their own reactivity vis-à-vis the predominantly Western anti-whaling sentiments. Yet both the primary and the secondary data provide plenty of examples of how this manifests itself in the whaling controversy.

A piece of evidence that the actors in the Japanese whaling scene view themselves through the prism of what the opposite, anti-whaling side says about them, and how the anti-whalers behave is the concept of *han-han hogei* or *anti-anti-whaling*, that was identified in the emic perspective during this research. The first time *han-han hogei* came to my attention, it was casually mentioned during one of the whaling-related events at the beginning of this study. It was said half-jokingly, and it seemed that it was an ad-hoc definition of the current mission of the Japanese whaling community. However, as time went by, I heard more and more participants using the same words to describe themselves as a group and the beliefs that were guiding their work. There is, in fact, no Japanese equivalent for "pro-sustainable whaling" that would be used by all of the representatives of Japanese whaling. People would use longer descriptive or more specific words to talk about it, or, in most cases, say *han-han hogei.*

The convenience of using "anti-anti-whaling" for the emic side can be tied to what was already discussed in the previous section – the normalcy of whaling for most Japanese people, and especially for those who were involved in it before the moratorium came into effect, or whose families and communities were involved in it for generations. As established earlier in this chapter, even now, with all the attention the problem has received in the media in the last few decades and with popularization of the anti-whaling rhetoric through Western pop-culture, the social landscape of Japanese whaling still has weak anti-whaling presence, that in essence is just non-pro-whaling rather than anti-. This was confirmed during many of the conversations I had with my participants. As one staff member from the WAO, Fisheries Agency of Japan, Ministry of Agriculture, Forestry and Fisheries shared the following:

> Japanese people do not think that there is anything wrong with eating whale. They can say there is a problem with research whaling, or there is no point going all the way to Antarctica using government funds, there are quite a lot of people who think that it is better to just do coastal whaling. But I think there are almost no people who would say we should not take even one whale.
>
> (personal communication, July 2017)[45]

If this is the current situation, it is easy to imagine that in the 1970s, when the anti-whaling campaigns started to appear in the West accompanying a broader move-ment of environmental consciousness, people in Japan were even less critical of whaling. They were not pro-whaling, they were simply taking it for granted. Given that eating whale meat was a norm in Japan, it took norm contestation from the outside for some local actors to reflect on it and formulate their own opinions. Un-derstandably, there cannot be any *pro-* without there first appearing some form of *anti-*. Giving a simpler food example, there is no pro-cucumber movement because we have plenty of them available, their price is attractive, and no one has strong reservations about other people eating cucumbers. Cucumber eaters do not have to justify it, explain themselves, or talk other people into also eating cucumbers be-cause it's, say, a tradition. They are not pro-cucumber then, their position is neutral.

And that is exactly the position most current anti-anti-whalers were starting from – neutral. This does not mean that it was a good place to be in given the col-lapsing whale stocks back at that time. But a lot of what the Japanese people and later their foreign allies have been doing starting approximately from the 1970s was a manifestation of reactivity. *Han-han hogei* is a linguistic reflection of that.

To phrase it using the conceptual framework of this study, the etic perspective has been of crucial importance in shaping the Japanese emic.

When having conversations in the field, many of my respondents would start explaining Japan's position, as well as their own position within the national discourse, through negation of what the anti-whaling forces were alleging. These monologues would often turn to the "they say that…but we are not like that at all" model, demonstrating the participants' remarkable awareness of the anti-whaling discourse and the defensiveness of the Japanese counter-stance.

However, reactivity is not only valid as a concept for understanding personal and group experiences. It is equally useful when it comes to explaining the concrete actions of the Japanese whaling actors. That dimension of reactivity is a lot simpler to see and assess. A lot of work was done by the Japanese whaling community in direct response to the actions of the anti-whaling forces.

On the national government level, such an effect of anti-whaling resulted in a waste of resources. In Western discourse it is common to view animal rights and environmental NGOs as entities who either bring about positive changes or, in the worst-case scenario, fail to make a difference. The case of Japanese whaling, however, showcases how anti-whaling NGOs involved in this matter had a negative overall effect. One example is that their campaigns increased the workload of government officials. From the testimonies of people serving at the WAO, some of them were focused exclusively on protecting the Japanese research whaling fleet from the activities of Sea Shepherd in the Antarctic Ocean in 2008–2015. For instance, one officer at the WAO shared the following:

> Since I got here in summer last year until this March I was mostly doing work related to the security [of the Japanese whaling fleet] in relation to Sea Shepherd… I cannot share too many details about it, but for example dispatching an extra ship for the research whaling fleet to be safer at sea and all the related documentation and consultations with other departments in the government was a part of what I was working on.
>
> (personal communication, July 2017)

The dispatching of an additional vessel mentioned in the above quote brings us to the second point. Because of the reaction to anti-whaling efforts, the cost of Japanese research whaling went up considerably. The cost embedded in the workload of the people at the WAO is a part of the increased financial burden. The extra equipment acquired for protection, as well as chartering one more vessel exclusively because of the threat Sea Shepherd posed added up to an even bigger difference in spending. Ironically, anti-whalers then would also accuse the Japanese government of spending too much public funds on research whaling, failing to recognize that some of it is a direct consequence of anti-whaling actions.

Another WAO staff member commented on the situation as follows:

> I think this criticism [excessive spending on whaling] is absolutely not true. Originally whaling did not have much budget, before around 2007–2008 it

almost did not cost any money at all. After that it became larger, and the primary reason for that is sabotage (bougai koui 妨害行為). Additional staff was commissioned to work in the whaling office at MAFF, an additional ship was engaged in the whaling operations, and we had to acquire equipment that we would not otherwise use, such as water cannons. An extremely large amount of money has been used for all that. Additionally, some money is being used after the ICJ case for the adjustment of the research whaling program in accordance with the court's decision. Otherwise, before that whaling was subsidized, but the money was coming back. So it was not something that required a lot of money. When the budget will be revised in due time if the costs go down, naturally the budget will become lower too. If the activities of Sea Shepherd stop, the costs will immediately go down.

(personal communication, July 2017)

When it comes to the process anti-whalers followed, not only did it not help them achieve their ultimate goal – since Japan still continues whaling, it also made the Japanese whaling community skeptical of ENGOs in general. Many of my research participants voiced feelings of distrust and contempt toward the environmental movement, extrapolating their judgment of the anti-whaling groups, who they see as primarily interested in attracting donations, to all those engaged in animal and nature activism. Some of the representatives of the Japanese whaling community often used air quotes when saying "NGOs". An experience one of the research participants shared on his encounters with the Sea Shepherd Conservation Society that exemplifies a general sentiment of the whaling community for anti-whaling activists can be summarized as "this is a performance". Consider this quote:

For example, they are letting us take whales. Meaning that they want scenes when we actually catch it or cut the whale body, when there is for sure blood. They want this kind of videos. It is rare for them to come before we catch a whale and prevent us from killing it. As if they were saying, please, take as much as you need, this will look more interesting on camera. I saw that while working on the ship, and that made me think that they are not that serious, that this is a performance.

(personal communication, September 2017)

Detrimental effects of aggressive anti-whaling campaigning were also seen on the local level. Specifically, in the whaling town of Taiji in Wakayama Prefecture. This small and quiet place became well-known all over the world after a documentary called "The Cove" came out in 2009 "exposing" Taiji's dolphin hunts (Psihoyos, 2009). The graphic footage of a bloody cove, the dialogue that was carefully edited to make the Japanese fishermen seem unwitty and cruel, and the mystery-drama style narrative showing the protagonists as brave and adventurous in the face of the "danger" posed by the locals won the hearts of audiences as well as several prestigious film awards. What stayed behind the scenes was the harassment of the townsmen to provoke negative reactions that could be then filmed on camera, as well as the life in Taiji post-"The Cove".

The town received unprecedented attention from both national and international media, as well as NGOs, whose representatives would camp on the town's streets for weeks during the hunting season. Animal rights and environmental NGOs saw a big opportunity to protest specifically in Taiji, as the documentary provided so much public exposure to this place. According to the mayor of Taiji – Kazutaka Sangen – and the Deputy Chief of the General Affairs Department of Taiji town government – Masaki Wada, during the first several months after the film came out there were incessant calls and messages from everywhere across Japan, as well as from overseas. Some of the callers were threatening the public servants and the town's community (personal communication, July 2017). The locals reported feeling vulnerable and misunderstood. The situation was aggravated by the fact that most anti-whalers were foreigners – they did not speak any Japanese and did not try to communicate with Taiji residents. In response to these anti-whaling activities, the local police was reinforced with officers from Wakayama city.

Two field trips to Taiji were conducted during this research – in 2016 and 2017, both of them showing that anti-whaling activists did not achieve their goals here. During my visits, there were only a few people representing this group outside the cove, where dolphin hunts still occur. The momentum was lost and most people who were[46] against the dolphin hunts, left Taiji. The reactivity of Taiji in this case was to mobilize in the face of an external threat. From the conversations with locals, it became clear that this situation only made them more determined to continue whaling. With the looming danger of losing this source of food, income, and identity, they reconfirmed their strong whaling traditions. As a result – whaling/dolphin hunts still occur in Taiji according to the government-issued permits, considerable resources were wasted to protect the town and its people from intruders, and the locals are now left suspicious of international influences in general and animal protection movements in particular.

These examples demonstrate that the Japanese reactivity in whaling should be considered, not only from a theoretical perspective but also purely practical. Anti-whaling forces should be more aware of *han-han hogei* in their assessments of the possible future steps in the whaling debates if they want to avoid counter-productivity and unnecessary damage.

It is interesting to continue following the developments in whaling now that the country is no longer a member of the IWC and has discontinued the most contentious component of its whaling policy – the research programs in the Southern Ocean. So far, according to a communications specialist with the ICR in Tokyo, *"suddenly everything stopped"*, meaning that the strong media attention to Japanese whaling is no longer there after the country withdrew from the IWC and discontinued its research programs in the high seas. The person also acknowledged that *"there is still some noise from the most vocal [anti-whaling] entities"*, but it's nowhere near the pre-2019 levels (personal communication, March 2023).

Non-pro-whaling and *anti-anti-whaling* show that the simpler pro- and anti-, as most dichotomies, are not sufficient to explain the more nuanced and complex reality. The whaling debates are not as clearly divided into the "good" anti-whalers and the "bad" pro-whalers as it seems to an untrained media consumer's eye.

Orientation toward the future

In the international English-language publications pertaining to the topic of Japanese whaling, the focus is often on the pro-whaling (or as we now know, the "anti-anti-whaling") fight Japan has put up in the international arena. It might seem like all Japan does is "defying the world" with its research whaling (Whiteman, 2015), including fighting in the ICJ because of it. In other words, manifestations of *reactivity* in Japanese whaling are highlighted, because this is the part of it that directly relates to the West and its opinions, therefore is of more interest to Western readers and viewers. In this book, reactivity is also given due attention and recognized as one of the pillars of Japanese whaling. However, there is another dimension to whaling policies and practice in Japan that has also heavily influenced the development of the issue. This dimension is self-reflective and self-aware, it is embedded in experiencing[47] the now of Japanese whaling through the lens of its ideal future. Some of the participants of this study – institutions as well as individuals – saw the goal of their work in the resumption of commercial whaling. But most of them also recognized verbally or demonstrated with their actions that *han-han-hogei* is an interim objective, and more work is needed in the domestic whaling-promotion arena. This whole debate would have no meaning if having won, Japan would find that it still lost because the timing was lost. What the actors in the Japanese whaling scene are most concerned about, is the irreversible loss of consumer interest, both in Japan and in the countries that supported Japan in the IWC. This idea was clearly voiced after Japan announced its withdrawal from the organization. First at a MAFF whaling discussion session that gathered most of the various whaling-related people on January 21, 2019. Later this was repeated in a conference room in the headquarters of the Liberal-Democratic Party of Japan, where some of the same people, as well as journalists, were treated to the leading party's take on the latest developments in whaling on February 1, 2019. Multiple speakers stated that no one should forget that "*this is not the end, this is just the beginning*" (personal observations). Meaning that the Japanese whaling complex should look into the future and be guided by an ambitious and transformational vision of itself. In 2023 I reconnected with some of the participants of this research, and they confirmed that the goal of reviving the domestic market was still on everyone's mind four commercial whaling seasons after Japan left the IWC.

Japan's self-sufficiency ratio

Ambitious is not an understatement when it comes to the Japanese whaling community sharing ideas of why whaling should and could become more important for their country. The civil servants at the WAO, MAFF, representatives of the ICR, representatives of the JWA, the founders of the WFF, restaurant owners, and university professors – all voiced the opinion that whaling has the potential to improve Japan's self-sufficiency ratio. This aspiration is an integral part of the shared emic perspective on Japanese whaling.

A country's food self-sufficiency ratio is defined as a percentage of domestic consumption to domestic production. Japan has been struggling with this index for decades. It is considered to be low at 37% in fiscal 2020[48] and the country consistently falls short of its target of 45% (Jiji, 2017). Whaling, if revived on a commercial scale, is seen as a potential contributor to this problem's solution, as it could help diversify the country's own food base. This was explained by Joji Morishita – a Tokyo University of Marine Science and Technology professor and former IWC Commissioner and Japan's delegation representative – as follows:

> In my mind this issue is related to food security. I often use a quote from Jacques Diouf – the Director-General of the United Nations' Food and Agriculture Organization[49]. He made a statement in 2006 saying that currently more than 90% of human population lives on 23 different food sources – 15 plant and eight animal. And that is it. Only 23 materials are feeding more than 90% of us. This situation is scary. You only have that much food in your food basket. There are diseases and droughts. If something happens, we might lose one of the 8. And at the same time, probably less than 10 countries among the 200 countries of the world can have a surplus, or a capacity of exporting food. Resilience, or food security, depends on diversity, or biodiversity. Japan has a very low self-sufficiency rate, only 39%[50]. Many Japanese nationals might not be aware of that. But 39% is a very scary figure. If imports stop, almost 2/3 of the Japanese population will have nothing to eat. We have been depending on the US and China for I think around 45% of our imports. All those facts considered, whaling is a small issue, but still a very important one.
>
> (personal communication, November 2016)

Morishita is a well-known figure in Japanese whaling and often speaks at whaling-related events, so some other community members might have first heard this explanation from him, and/or he might have heard it from his colleagues or predecessors. The ties between the various elements of the social landscape of Japanese whaling are so numerous that it would be now hard to tell where the self-sufficiency ratio idea originated and how it spread. But it is the most readily shared explanation of where the Japanese whaling community wants to take this industry in the future. It is not only a part of the official government discourse but also what people on all levels of involvement tell each other during the most private gatherings.

From the etic perspective, taking into consideration the current lack of popularity of whale meat on the Japanese market this aspiration sounds unrealistic. However, with Japan having withdrawn from the IWC and restarting commercial whaling in 2019, the figures might show a different tendency in the near future. The passion and determination of the representatives of the Japanese whaling community were apparent throughout this study – now the question is whether it will be sufficient to turn their hopes into reality.

For now the methods employed for this idea to materialize are mostly related to whaling-related education and promotion of whale-based cuisine. As soon as one starts following the key people in this domain, it becomes clear that promotional whaling activities are organized frequently and with relative success. Some of them are small, inviting 10–15 people, but some attract up to 500 attendees. I asked a representative of the JWA to compile a provisional list of the whaling-related events supported by the JWA in the 2017–2018 fiscal year,[51] and that unofficial document included 35 events. The JWA does not coordinate all such activities, and some are organized independently in small towns, or by other whaling-related entities, so this number reflects only a part of what is out there. I attended only select events during this project, and even my personal count was close to 30 between 2016 and 2018.

The whaling industry, like any other commercial activity, was not able to develop at the expected pace during the COVID-19 years, and in 2023 Japan is arguably still affected by the pandemic.[52] However, the whaling community has found a way to adapt and persevere. For example, some whaling-related events were held online by the WFF, and Koydo Senpaku, under the guidance of their new president Hideki Tokoro, started to sell whale meat in vending machines (personal communication, March 2023). With more freedom for promotional campaigns as whaling became a regular commercial activity from July 2019 and as the country emerges from the COVID-19 pandemic, it is possible that these efforts will grow in scale and frequency.

International pro-sustainable whaling leadership

During one of our conversations, Joji Morishita – who has been involved in the international and national negotiations concerning Japanese whaling for more than 20 years, including in the role of Japan's commissioner to the IWC, then IWC chair in 2016–2018 – noted that at some point it was decided to not use the "culture argument" anymore (personal communication, December 2016). The reason for this decision is related to Japan finding itself in the position of international leadership when it comes to pro-sustainable whaling. As discussed earlier in this chapter, Japan still finds valid the argument that culture is reason enough to keep whaling, but it also supports a position that a whaling industry could also be developed in countries that do not have historical or cultural ties to this practice, as long as sustainability is embedded in the strategy. The small island nations, as well as West African countries that were Japan's allies when it was a member of the IWC, could potentially become whaling nations in Japan's opinion. While the idea outrages the West, if we suspend our judgment and imagine it to be any other industry, this could be a reasonable proposition in certain cases. There are many areas where populations struggle with food security and at the same time have strong bonds with the ocean and are used to receiving their calories from marine resources – developing whaling industries here could succeed.

Currently there are over 40 countries members of the IWC who officially support the position that whales are a consumable resource. For instance, consider these comments given by some delegations during the process of voting on a Schedule Amendment on Aboriginal Subsistence Whaling[53] at IWC67, held in Florianopolis, Brazil in 2018:

> We think that there is no need to justify it as subsistence [whaling], as long as sustainability is guaranteed.
>
> (a representative of Solomon Islands)

Nutrition and food security are basic human rights. This is scandalous and representative of the dysfunctional nature of this organization. [community] Needs should only be discussed for endangered species.

(a representative of Antigua and Barbuda)

The Scientific Committee recommendations are supportive [of the proposal]. We think that discussing [community] need is intruding on a nation's right to self-determination. It is unnecessary to talk about the need [in this context]

(a representative of Norway)

There is no need to talk about [community] needs, simply about an adequate procedure to guarantee stocks' health. It is interesting that countries like Chile have a problem with automatic renewal [of catch quotas], in light of the moratorium.[54]

(a representative of Iceland)

(all of the above quotes are from personal observations, September 2018)

These prove that when it comes to whaling, certain countries strongly believe in rooting their decisions on the science of wildlife resource management rather than on the perceptions about wildlife. It is this camp that Japan was leading and can still lead after its withdrawal from the IWC.

To make this goal a reality, several measures were taken by Tokyo. For instance, representatives of the ruling Liberal Democratic Party of Japan personally visited each of the whaling-friendly countries right before or soon after December 26, 2018 (the date of the official announcement of Japan's withdrawal from the ICRW/IWC) in order to explain Japan's decision and assure of Tokyo's intention to not give up its leadership position in the area of sustainable whaling. This is a significant diplomatic move that shows Japan's determination to sustain and further develop its relations with its former IWC allies. The countries-allies list included Guinea, Tanzania, Kenya, Vietnam, Laos, Cambodia, Norway, Iceland, Denmark, Antigua and Barbuda, Saint Lucia, Saint Vincent, and the Grenadines. Reports on these visits were made at the LDP post-IWC withdrawal gathering on February 1, 2019 (personal observations).

Post-IWC withdrawal, Japan is still in close communication with other whaling nations. For instance, in Denmark's Faroe Islands, where the town of Klaksvik has whaling practices similar to those in Japan's Taiji – drive hunting. In 2018 Taiji and Klaksvik officially established a sister-cities relationship, with both mayors visiting each other's towns and discussing the prospects of future professional exchanges. This relationship is currently going strong.

Another case of Japan maintaining close ties with its allies is its relationship with the whaling industry in Iceland. Notably, in 2022 Iceland's Minister of Fisheries and Agriculture made remarks indicating that whaling might be discontinued in this country from 2024, when the current catch quotas are set to expire

(BBC, 2022). Like Japan, Iceland is no stranger to continuous criticisms from anti-whaling organizations and activists that put pressure on its tourism and reputation in general. However, it remains to be seen if the country proceeds with the idea of ending whaling forever. If Iceland's whaling industry does cease to exist, Japan's international position could become even trickier with no Western countries backing it up. For now, however, there is one whaling company operating in Iceland – Hvalur hf. and its CEO is of the opinion that whaling is a viable and sustainable industry. In 2023 2700 tons of Icelandic whale meat was exported to Japan, to be distributed by the Japanese Kyodo Senpaku. That amount is larger than what was sold by Kyodo Senpaku in 2022 – about 1800 tons (personal communication, March 2023), which demonstrates both the company's ambitions with regard to developing its domestic market and its willingness to support the whaling industries abroad in line with its leadership position in this arena.

Interestingly, regardless of the fact that Japan is no longer a member of the IWC, the country still sent its delegation to the 2022 IWC Committee Meeting – the first one held post-COVID pandemic. Joining in the status of an observer this time, Japan signaled its continued interest in the issue and acted on the statement made by the then Prime Minister Abe's cabinet, saying that the government of Japan "will continue to contribute to the science-based sustainable management of whale resources" (2018).

Notes

1 The whole time Japan was a member of the IWC its propositions were formally given time and space. But many of the participants of this study, who were regular attendees of the IWC Committee meetings voiced frustration at how the decisions were pre-determined for the anti-whaling side and it never mattered what arguments and evidence was presented by Japan and its allies (personal communication, 2016–2019). Similar opinions were shared with me during the IWC67 Committee meeting in Brazil in September 2018 – the last one the Japanese delegation attended as a member of the organization.

2 This description of all cetaceans regardless of the species can be found in a variety of popular culture and media sources, as well as anti-whaling speeches and remarks. See, for example, the tweet of Great Britain's environment secretary posted in reaction to Japan's announcement of its withdrawal from the IWC: "Extremely disappointed to hear that Japan has decided to withdraw from the IWC to resume commercial whaling. The UK is strongly opposed to commercial whaling and will continue to fight for the protection and welfare of these majestic mammals" (quoted in McCurry & Weaver, 2018).

3 This particular word combination is from a quote attributed to "conservation groups" in a The Guardian article by McCurry & Weaver (2018). It captures the views of the anti-whaling forces in general.

4 This idea with some variations was expressed throughout the book by Kalland and Moeran (1992), as well as several of this study's participants (personal communication, 2016–2019).

5 It is in reality not "the rest", as demonstrated by the number of pro-whaling votes at the IWC that at times reaches the majority depending on the issue (e.g., St. Kitts and Nevis Declaration of 2006, available online https://archive.iwc.int/pages/download.php?ref=2081&size=&ext=pdf&k=&alternative=-1&usage=-1&usagecomment=), and the IWC discussions, in which pro-sustainable whaling countries' delegates actively participate (personal observations, September 2018). There are countries besides Japan that defend pro-sustainable whaling policies, but their opinions are rarely highlighted in the media.

6 See Chapter 1 on the threat of the US sanctions against Japan and the effect this had in the wake of the IWC-imposed moratorium on all commercial whaling.

7 Anti-whaling activists have taken to the streets ever since the inception of environmentalism in the 1970s. Certain events and actions of the Japanese side provoked peaks in such activities throughout the past five decades, but there were also relatively calm years for pro-sustainable whalers. I witnessed peaceful demonstrations of the representatives of "Sea Shepherd Conservation Society", some other organizations and independent activists in the airport and around the venue of the IWC67 Committee meeting in Florianopolis, Brazil in September, 2018. Another demonstration was organized on January 26, 2019 in London in protest of Japan's decision to withdraw from the IWC and resume commercial whaling. See a video from the march to the Japanese embassy in London here https://www.youtube.com/watch?v=-Q98Fgl2beE

8 See footnote 18 for information on the 2008-2015 Animal Planet show "Whale Wars".

9 There are numerous examples of anti-whaling activists using derogatory slurs against people connected to Japanese whaling. See, for example, a blog post by the founder of "Sea Shepherd Conservation Society" called "Meet Ginza Glen – a corporate whore for the whale killers" (2007), where besides calling Glenn Inwood (his first name is misspelled throughout the text of the blog post) names and ridiculing his work, it is alleged that he has betrayed his own country of New Zealand by working for the Japanese government and suggested that readers call or mail him directly to express any negative feelings they have for him followed by sharing his contact details. Ironically, the very same author often appeals to the moral and ethics of people who support whaling. See the post online https://www.seashepherd.org.au/latest-news/meet-ginza-glen-corporate-whore-for-the-whale-kill/

10 See more on the concept of super-whale and its functioning within the whaling debate further in this chapter.

11 See, for example, "Kujira" by Danny Samit (2008), "A Whale of a Tale" by Megumi Sasaki (Psihoyos, 2016).

12 See, for example, a TED talk by Simon Wearne, who was a former film crew member for the "Whale Wars" TV-show and now researches the history and traditions of whaling in Taiji https://www.youtube.com/watch?v=qmWc_Wvykzk; an interview with Matthew Barney – an artist, who found inspiration in traditional small-town Japanese whaling https://www.sfmoma.org/watch/matthew-barney-and-ancient-japanese-whaling-traditions/

13 See, for example, https://archive.vn/luna.pos.to - a blog featuring numerous whaling-related posts with a predominantly neutral way of presenting related information, including statistics translated from Japanese institutions' websites and articles by guest writers.

14 See, for example, Kalland and Moeran (1992).

15 The word "defy" is routinely used to describe Japan's actions in the context of whaling. See e.g. Whiteman (2015).

16 There is no official definition of "semi-academic", but there are now many online resources where academics, policy practitioners and other professionals of their fields publish opinion pieces that do not undergo the standard academic journals' peer-review process. This makes the "semi-academic" pieces valuable, but they are not vetted as rigorously as academic ones.

17 See, for example, Takahashi et al. (1989), Kalland & Moeran (1992). Although the title of the second work indicates a holistic approach – "Japanese Whaling: End of an Era?" – the main focus is on "whaling culture", as well as the development and transferring of whaling-related knowledge throughout the history of this practice. The political side of the issue is given less attention, not the least because at the time the research for this book was done the enmity between the anti- and pro-sustainable whaling camps had not yet reached its peak levels.

18 See, for example, Akimichi et al. (1988); Fisher (2016); Heazle (2004).

19 See, for example, Clapham et al. (2003); Clark & Lamberson (1982); Gales et al. (2005).

20 See, for example, Andresen (1993); Morishita & Goodman (2005); Morishita & Goodman (2011); Morishita (2023); Nagtzaam (2009).

21 See, for example, Nussbaum & Nussbaum (2016); Skodvin & Andresen (2003); Ruffle (2002).

22 See, for example, Hamazaki & Tanno (2001); Bowett & Hay (2009).

23 See, for example, Mangel (2016); McClintock (2017); Peel (2015).

24 See, for example, Kolmaš (2021); Normile (2019).

25 See, for example, a news report by Darby for The Sydney Morning Herald (2009); a news report by Willacy for a popular Australian broadcasting corporation ABC (2011); an academic article by Blok (2008) or one by Prys-Hansen et al. (2023).

26 Pre-IWC withdrawal, the Japanese delegation to the IWC used to be one of the biggest, comparable to the one representing the USA. This is related to the fact that Japan, unlike many other nations members of this "whaling commission", was still whaling, although not commercially and not for larger species of whales until 2019 when it left the IWC – smaller species not regulated by the IWC were continuously taken in accordance with Tokyo-issued quotas and research whaling was also being conducted. That is why the presence and active participation of the representatives of MAFF was required, while for other delegations no ministries' representatives other than the ones dealing with foreign affairs were necessarily included.

27 Since the positions of public officers within the ministries are rearranged every several years, which will be discussed further in this section, some of the people joined WAO closer to the end of this study and were not interviewed. By 2023, the head of WAO is also a different civil servant – the new head of the office did not participate in this study.

28 Whaling in the Antarctic (Australia v. Japan: New Zealand intervening), 2010–2014.

29 See the ICR about page here https://www.icrwhale.org/abouticr.html

30 Ibid.

31 The ICRW, Article VIII provides for members of the IWC to issue permits for the taking of whales for research purposes. Since Japan withdrew from the agreement, it had to stop its operations in the high seas.

32 See the website of the JWA, "About" section www.whaling.jp/english/intro.html

33 See the website of Kyodo Senpaku Kaisha Ltd.
    http://www.whaling.jp/intro.html (in Japanese)

34 Gaibo Hogei was mentioned in the context of whaling in Wadaura in Chiba prefecture – one of the important whaling towns I discuss throughout this book.

35 The majority of respondents had positive views on Tokyo's decisions to leave the IWC and resume commercial whaling – 67.7% chose "highly value them" or "somewhat highly value them". See the poll's results here https://www.mofa.go.jp/press/release/press4e_002430.html

36 See IKAN's website http://ika-net.jp/en/

37 See LIA's website https://ngo-lia.org/en_about.php

38 See, for example, the description of an anthropological research under the auspices of the National Museum of Ethnology (Osaka, Japan) on keeping whaling traditions in the context of anti-whaling activism http://www.minpaku.ac.jp/research/activity/project/other/kaken/15H02617; multiple articles on the "Japan Forward" portal, such as this one https://featured.japan-forward.com/whalingtoday/2022/12/01/nammco-at-30-marine-mammals-in-cultural-traditions-and-identity-part-2-of-3/

39 A video of the performance filmed by a festival attendee can be viewed here https://www.youtube.com/watch?v=jIBxtzkcgQE&t=33s

40 See the website here https://www.town.taiji.wakayama.jp/

41 See, for example, Wingfield-Hayes (2016).

42 Certain critics go as far as making comparisons of whaling with female genital mutilation – a painful and dangerous procedure that has been performed for non-medical reasons in a number of communities across the world. Anti-whaling supporters have

made the unlikely connection suggesting that both are traditions, but they do not belong in the modern-day world. The statement was made by the representatives of the Sea Shepherd Conservation Society, but the article containing that statement originally published on a New Zealand news portal is no longer available online. The same thoughts are expressed by, for example, Walker (1999). Some non-specialist bloggers, as well as commentators on whaling-related videos and articles also picked up on the comparison, demonstrating the ease with which such logically invalid, sensationalist statements can win popular support. See, for example, a blog post titled "Female circumcision = Whaling" (accessed online https://thegreatestperfect.weebly.com/meanderings/female-circumcision-whaling).

43 *"Kujira"* (formerly *"Himitsu Kujira"*) was opened in May 2015. Each time I visited it – thrice during this research – all of the tables and seats at the bar were occupied or reserved. When asked about how the restaurant was doing during my interview with him, two years after the place's opening Ishikawa shared the following: *"I think it is going in accordance with what we were aiming at. Of course, I would be happy if we had more guests [laughs].. But comparatively speaking we are doing well, on par with other restaurants in the same segment of the market. I think going on for two years with a menu based almost exclusively on whale meat is a really good result in itself."* (personal communication, May 2017).

44 This is not a direct quote from any one publication, but the general image perpetuated by the anti-whaling side of the debate.

45 This quote also shows that the government had been aware of people prioritizing STCW and had troubles sustaining Antarctic whaling and continuing the confrontation within the IWC. Japan's withdrawal from the organization was effectively executed by Japan almost exactly two years after this interview was conducted.

46 It is hard to tell whether the activists who spent time in Taiji for a few years after "The Cove" came out in 2009 are still interested in the issue, since it has not been resolved in their favor, but most of them seemingly lost interest.

47 "Experience" here is used in broad phenomenological sense.

48 The latest data on Japan's food self-sufficiency ratio is available on the website of the Ministry of Agriculture, Forestry and Fisheries https://www.maff.go.jp/e/data/publish/attach/pdf/index-69.pdf

49 Jacques Diouf (born August 1, 1938) is a Senegalese diplomat who was the Director-General of the United Nations' Food and Agriculture Organization (FAO) from January 1994 to December 2011.

50 At the time of the conversation this number was correct at 39%, according to the latest available data, it dropped 2 points by FY 2020. The latest data on Japan's food self-sufficiency ratio is available on the website of the Ministry of Agriculture, Forestry and Fisheries https://www.maff.go.jp/e/data/publish/attach/pdf/index-69.pdf

51 The concept of "whaling-related events" is a product of the present research, all of these projects are different in nature, scope and goals, so they were not previously analyzed as one category by JWA or any other whaling-related organization.

52 For example, March 13, 2023 was the first day when the government-issued indoor mask wearing guidance was relaxed in Japan.

53 The proposal was introduced by the USA with the co-proponents of Denmark, Russia and St Vincent and The Grenadines, who all currently have communities engaged in what is recognized as aboriginal subsistence whaling. The proposed amendment was adopted with 56 votes in favor. More details on this proposal can be found in IWC67 Report, p. 7–9. Available online https://archive.iwc.int/pages/download.php?direct=1&noattach=true&ref=7592&ext=pdf&k=

54 The speaker alludes to the fact that the moratorium on commercial whaling was originally introduced in 1982 as a temporary measure, but it was never formally reassessed and hence was *de facto* renewed automatically.

References

Akimichi, T., Befu, H., Braund, S. R., Hardacre, H., Kalland, A., Moeran, B. D., & Takahashi, J. (1988). *Small-type coastal whaling in Japan: Report of an international workshop.* Edmonton: Boreal Institute for Northern Studies.

Andresen, S. (1993). The effectiveness of the International Whaling Commission. *Arctic, 46*(2), 108–115.

BBC (2022, February 4). Iceland whaling: Fisheries minister signals end from 2024. https://www.bbc.com/news/world-europe-60265085

Blok, A. (2008). Contesting global norms: Politics of identity in Japanese pro whaling countermobilization. *Global Environmental Politics, 8*(2), 39–66.

Bowett, J., & Hay, P. (2009). Whaling and its controversies: Examining the attitudes of Japan's youth. *Marine Policy, 33*(5), 775–783.

Carney, M. (2014, May 09). Japan restarts whaling program in north-west Pacific from base at port of Ayukawa. *ABC Premium News.* Accessed online https://www.abc.net.au/news/2014-05-09/japan-continues-whaling-in-north-west-pacific/5441426

Clapham, P. J., Berggren, P., Childerhouse, S., Friday, N. A., Kasuya, T., Kell, L., & Brownell, R. L. Jr. (2003). Whaling as science. *BioScience, 53*(3), 210–212.

Clark, C. W., & Lamberson, R. (1982). An economic history and analysis of pelagic whaling. *Marine Policy, 6*(2), 103–120.

Darby, A. (2009, May 4). Japan won't budge on whale kill. *The Sydney Morning Herald.* https://www.smh.com.au/environment/conservation/japan-wont-budge-on-whale-kill-20090503-arft.html

Fackler, M. (2011, March 24. Japanese town mulls future without whaling industry. *The New York Times.* Accessed online https://www.nytimes.com/2011/03/25/world/asia/25whale.html

Fisher, S. (2016). Japanese Small Type Coastal Whaling. *Frontiers in Marine Science, 3*(121), 1–6.

French, D. P., & Sutton, S. (2011). Does measuring people change them? *Methods, 24*(4), 272–274.

Gales, N., Kasuya, T., & Clapham, P., & Brownell, R. (2005). Japan's whaling plan under scrutiny. *Nature* 435, 883–884.

Hamazaki, T., & Tanno, D. (2001). Approval of whaling and whaling-related beliefs: Public opinion in whaling and non-whaling countries. *Human Dimensions of Wildlife, 6,* 131–144.

Heazle, M. (2004). Scientific uncertainty and the international whaling commission: An alternative perspective on the use of science in policy making. *Marine Policy, 28*(5), 361–374.

Kalland, A., & Moeran, B. (1992). *Japanese Whaling: End of an era?* Copenhagen: Scandinavian Institute of Asian Studies.

Kolmaš, M. (2021). International pressure and Japanese withdrawal from the international whaling commission: When shaming fails. *Australian Journal of International Affairs, 75*(2), 197–216.

Kumano Shimbun (2018, May 1). Kujira no onkei ni kansha – Kandorizaki koen de kuyosai – Taijicho [Gratitude to the blessing of whales – Memorial service at Kandorizaki Park – Taiji town]. Accessed online https://kumanoshimbun.com/press/cgi-bin/userinterface/searchpage.cgi?target=201805011101

Mangel, M. (2016). Whales, science and scientific whaling in the international court of justice. PNAS. Accessed online https://www.pnas.org/content/113/51/14523

McClintock, C. C. (2017). Greasy luck to whalers: How the International Whaling Commission and International Court of Justice can use principles of American administrative and environmental law to keep Japan from circumventing the International Convention for the Regulation of Whaling. *Chicago Unbound*, International Immersion Program Papers, 53. Accessed online https://chicagounbound.uchicago.edu/cgi/viewcontent.cgi?article=1052 &context=international_immersion_program_papers

McCurry, J., & Weaver, M. (2018, December 2018). Japan confirms it will quit IWC to resume commercial whaling. *The Guardian*. Accessed online https://www.the-guardian.com/environment/2018/dec/26/japan-confirms-it-will-quit-iwc-to-resume-commercial-whaling

Megumi, S. (Director). (2016). *A Whale of a Tale* [Film]. Fine Line Media.

Morishita, J. (2023). IWC68: Reflections on the Future of the International Whaling Commission. *Whaling today*. Accessed online https://featured.japan-forward.com/ whalingtoday/2023/01/13/iwc68-reflections-on-the-future-of-the-international-whaling-commission/

Morishita, J., & Goodman, D. (2005). Role and problems of the scientific committee of the international whaling commission in terms of conservation and sustainable utilization of whale stocks. *Global Environmental Research*, *9*(2), 157–166.

Morishita, J., & Goodman, D. (2011). The IWC moratorium on commercial whaling was not a value judgment and was not intended as a permanent prohibition. *Aegean Review of the Law of the Sea and Maritime Law*, *1*, 301–311.

Nagtzaam, G. J. (2009). The international whaling commission and the elusive great whale of preservationism. *William & Mary Environmental Law and Policy Review*, *33*(2), 375–447.

Normile, D. (2019, January 10). Why Japan's exit from international whaling treaty may actually benefit whales. *Science*. Accessed online https://www.science.org/content/article/ why-japan-s-exit-international-whaling-treaty-may-actually-benefit-whales

Nussbaum, W., & Nussbaum, M. (2016). The legal status of whales: Capabilities, entitlements and culture. *Sequencia (Florianopolis)*, *72*, 19–40.

Peel, J. (2015). Whaling in the Antarctic (Australia v. Japan: New Zealand intervening) (I.C.J.). *International Legal Materials*, *54*(1), 1–52.

Prys-Hansen, M., Burilkov, A., & Kolmaš, M. (2023). Regional powers and the politics of scale. *International Politics*. Accessed online https://doi.org/10.1057/s41311-023-00462-8

Psihoyos, L. (Director). (2009). *The Cove* [Film]. Diamond DocsFish, FilmsOceanic, Preservation Society.

Ruffle, A. M. (2002). Resurrecting the international whaling commission: Suggestions to strengthen the conservation effort. *Brooklyn Journal of International Law*, *27*(2), 639–671.

Samit, D. (Director). (2008). *Kujira* [Film]. Single Serving Films.

Skodvin, T., & Andresen, S. (2003). Nonstate influence in the international whaling commission, 1970–1990. *Global Environmental Politics*, *3*(4), 61–86.

Taiji Town (2015a). Ibento joho [Ivent information]. Taijicho- kuroshio to kujira koshiki hogei hassho no machi Taijicho kanko joho [Taiji town – sightseeing information for Taiji town where Kuroshio Current flows and the old-style whaling was started]. Taiji: Taiji Town.

Taiji Town (2015b). Taiji's cultural heritage. Taiji no isan. Taiji: Taiji Town.

Takahashi, J., Kalland, A., Moeran, B., & Bestor, T. C. (1989). Japanese whaling culture: Continuities and diversities. *Maritime Anthropological Studies*, *2*(2), 105–133.

Tani, K. (2012). Shouchiikibetsu ni mita hiashi nihon daishinsai hisaichi ni okeru shibousha oyobi shibouritsu no bunpu [distribution of the number of deaths and mortality rate in the areas affected by the great East Japan earthquake by region]. *Saitama University, Faculty of Education, Department of Geography, 32,* 1–26.

The Japan Times (2011, March 18). Miyagi coastal whaling port pulverized, little more than memory. Accessed online https://www.japantimes.co.jp/news/2011/03/18/national/miyagi-coastal-whaling-port-pulverized-little-more-than-memory/#.XFxiUKpKiUk

Wakayama Hoso Nyusu (2018, April 29). Kujira no megumi ni kansha, Taijicho de Kuyosai [Gratitude to the blessings of whales, Memorial service in Taiji town].

Walker, P. (1999). Makah whaling is also a political issue. *The Ecotone,* Fall, 8-10.

Whiteman, H. (2015, December 1). Japan defies world as 'research' ship embarks on minke whale kill. CNN. Accessed online https://www.cnn.com/2015/11/30/asia/japan-whaling-research/index.html

Willacy, M. (2011, August 11). Japan's fisheries float idea of ending whaling. *ABC.* Accessed online https://www.abc.net.au/radio/programs/am/japans-fisheries-float-idea-of-ending-whaling/2836332

Wingfield-Hayes, R. (2016, February 8). Japan and the whale. BBC. Accessed online https://www.bbc.com/news/world-asia-35397749

Conclusion

Japanese whaling for the country's whaling community and beyond

The concepts of "emic" and "etic" were central to this research. Emic being the view from inside, and etic being the outside perspective, these two components of understanding the whaling dilemma are equally important. These terms with origins in anthropology, and sociology, and rooted in these disciplines' fieldwork methodology, presuppose that "inside" means from the inside of a social group of interest, while "outside" is what a researcher brings into the study with them, through their own prior experiences and knowledge. In this study, the emic approach was inquired into as prescribed – accounts of own experiences were sought from the Japanese whaling-related individuals and observations of their whaling-related activities were conducted to better understand the internal elements and functioning of the Japanese whaling landscape. The concept of etic, however, was understood and utilized more broadly than the anthropological tradition proposes. Rather than the researcher's view, the views of the anti-whaling camp openly expressed in published and broadcasted resources were seen as the embodiment of the etic perspective. It was not as far a departure from the origins of the concept – after all, I, the researcher, also used to espouse the anti-whaling rhetoric. Emblematically of the problem, it was not because I had known much about it or cared a lot, but because it was the accepted default.

The relationship between anti- and pro-, or the etic and the emic if we look at it through the anthropological lens, is often seen as a simple antagonism in the whaling debates. In reality, this relationship is intersubjective. The anti- and pro- evolve together in a common environment and devise their next steps as responses to the steps of the opponent. They even use the same terms to justify their stance, which is a rarely noticed commonality between the two camps – some examples of that are the use of the "sustainability" concept by both pro-whaling and anti-whaling sides; or the discourse around working for the sake of future generations. It is also argued that the lack of understanding and respect toward the emic perspective of the opposite side of the pro- and anti-whaling spectrum is the cornerstone of the stalemate we currently witness in the development of the issue.

My initial ignorance about this issue, and, later on, the surprise of the whaling-related community at my undertaking of this topic, are some of the circumstantial pieces of evidence that point to the fact that whaling is not a "hot" topic these days. Popular media does cover whaling, but this coverage comes in waves, briefly

DOI: 10.4324/9781003255031-7

peaking when the Japanese whaling fleet starts its seasonal operations and ends them several months later. There are also spikes in attention to the matter every time it takes an unexpected turn, such as after the International Court of Justice's (ICJ) decision to not endorse Japan's research whaling program in the Antarctic in 2014, or, most recently, Japan's withdrawal from the International Convention on the Regulation of Whaling and the International Whaling Commission (IWC) in 2019. As recently as in the early 2000s the IWC's Committee meetings used to attract hundreds of media representatives, in the past few years this number has seen a dramatic decrease. At the 2018 IWC67 Committee meeting that I had a chance to attend, there were around fifty-five accredited media, with more than a third of them representing Japan and only a handful actually in sight during the deliberations. As the former Japan's Commissioner to the IWC and its Chair in 2016–2018, Joji Morishita noted comparing whaling to other environmental issues – "it is no climate change-level concern" (personal communication, November 2016). Many others agree with this assessment. Yet, after having conducted this research, I firmly believe that there is more to the issue of whaling than meets the proverbial eye and it is important to continue this conversation.

During this study, with the support and help from its participants, I discovered the connection of whaling to some significant issues that might not be immediately associated with it. For instance, Japanese government officials, as well as representatives of NGOs and small-business owners talked about improving the country's self-sufficiency rate, and possibly also improving this metric for other countries by developing a whaling industry for themselves. While sustainability-related calculations are complex, consuming sources of protein that are available within the country's territory (territorial waters in this case) and without having to use additional land for agriculture and pasture is something that should be at the very least given room for consideration.

Another issue that was often discussed is the negative role environmental non-profit organizations play in the global dynamics of whaling. While the positive effect of their work attracting the global community's attention to the drastic decline of whale populations in the 1970s is praised and recognized, the adverse outcomes of the environmentalists' activities are rarely discussed. Being able to view their activities with a critical eye is a crucial tool we as a society have in making these organizations more efficient and more accountable to their supporters.

One more important theme that emerged time and time again at different stages of this research is the unparalleled role of media in shaping and framing the processes related to whaling. The once-constructed image of Japanese whaling has been migrating from publication to publication, from video piece to video piece with almost no alterations for several decades now. The details are so ingrained in that picture through constant repetition of the exact same words and phrases, that it leaves no room for alternative views or explanations. This book argues that this process does not only influence the outside world's attitudes to Japanese whaling but also how the Japanese whaling-related people view themselves. Their self-definition is embedded in the discourse on whaling that Western media propagates, only it is realized through negation of it rather than confirmation. Thus, the particularities of the international problematization of whaling heavily impacted the way Japanese people view and conduct this activity. This aspect of the issue is given much less attention in both academic literature and popular media than the fact that the Japanese people refuse to completely give up on whaling and accept the new, supposedly global, anti-whaling norm.

The high importance of whaling for the representatives of the Japanese whaling-related community is manifested in their persuasion that what they do is "*shimei*" – their mission in life. This idea emerged in the shared narratives of this study's participants in explicit or implicit ways – both in the interviews and short conversations and in actions. The way people in this group go out of their way to participate in whaling-related events, including traveling to Tokyo from their remote towns and flying overseas to attend the International Committee meeting; the way they take on multiple roles related to whaling from cooking and hands-on work on board boats and ships to office work; and the way they insist on keeping this activity alive – it all shows that whaling is a big part of this community's identity. If that part disappears, it would definitely take many of them a while to reinvent themselves.

There is a general understanding that the issue of whaling is about "saving the whale". And this idea is in no way denied legitimacy in this book. The fact that whales, as much as other non-human animals and the environment overall, deserve our close attention and protection is unquestionably supported here. The work done to stop and reverse the loss of whale populations is a fascinating case study, providing an encouraging example for solving other environmental issues. We did drive certain species of whales close to extinction in the 20th century. But there is no denying that the story of saving "the whale" is also a positive one. There can be resonant claims that solutions were late to come, however, in critical times critical measures were in the end taken by the international community. This resulted in very promising tendencies in whale stocks' recovery – some species have been taken off the endangered species lists or moved to other categories of less pressing concern. And as such whaling and the response to it demonstrates how the global society can and should deliver.

But is this story over? And, more importantly for the main goal of this project, does this story end with "saving the whale"? The stance taken here is that saving the environment should go hand in hand with considering all changes in human conditions any such strategy entails. While in this particular study, the main topic is whaling and the Japanese people involved in the battle for their right to continue this activity, this research was not intended only as a statement in support of the Japanese side. Admittedly, after having listened to the many voices of Japanese whaling, it is difficult not to empathize with the humans who are in distress because of the current situation. Still, the overall aspiration of this undertaking was to broaden awareness of this issue and unpack the Japanese way of looking at the problem in order to be better informed as a global society and be able to make educated decisions in this and other cases with similar variables. Importantly, many other countries, mainly West African and Asian, share the Japanese vision of sustainable whaling as well – another fact worthy of more attention.

To put it simply, there can hardly be step-by-step instructions on how to resolve a situation when a community has been engaged in a certain activity for years, sometimes generations and the global society came to the conclusion that this activity is no longer (morally) acceptable. But respectfully and deeply inquiring into these people's experiences might help us to come up with better policies, better support initiatives, and ultimately, better outcomes for all parties.

Index